Stem Cells and Biomaterials for Regenerative Medicine

Stem Cells and Biomaterials for Regenerative Medicine

Edited by

Marek J. Łos

Andrzej Hudecki

Emilia Wiecheć

ELSEVIER

ACADEMIC PRESS

An imprint of Elsevier

Academic Press is an imprint of Elsevier
125 London Wall, London EC2Y 5AS, United Kingdom
525 B Street, Suite 1650, San Diego, CA 92101, United States
50 Hampshire Street, 5th Floor, Cambridge, MA 02139, United States
The Boulevard, Langford Lane, Kidlington, Oxford OX5 1GB, United Kingdom

Library of Congress Cataloging-in-Publication Data

A catalog record for this book is available from the Library of Congress

British Library Cataloguing-in-Publication Data

A catalogue record for this book is available from the British Library

ISBN 978-0-12-812258-7

Publisher: Andre Wolff
Acquisition Editor: Mica Haley
Editorial Project Manager: Sam Young
Production Project Manager: Punithavathy Govindaradjane
Designer: Greg Harris

Typeset by SPi Global, India

Contents

Contributors

Numbers in parentheses indicate the pages on which the authors' contributions begin.

Theresa D. Ahrens (17,37), Nordic European Molecular Laboratory (EMBL) Partnership, Centre for Molecular Medicine Norway, University of Oslo, Oslo, Norway

Shiva Akbari-Birgani (57), Faculty of Biological Sciences; Center for Research in Basic Sciences and Contemporary Technologies, Institute for Advanced Studies in Basic Sciences (IASBS), Zanjan, Iran

Hossein Ansari (57), Department of Medical Genetics, School of Medicine, Ahvaz Jundishapur University of Medical Sciences, Ahvaz, Iran

Karolina Bakalorz (1), Faculty of Chemistry, Silesian University of Technology, Gliwice, Poland

Maryam Tahmasebi Birgani (57), Department of Medical Genetics, School of Medicine; Cellular and Molecular Research Center, Ahvaz Jundishapur University of Medical Sciences, Ahvaz, Iran

Safak Caglayan (17,37), Nordic European Molecular Laboratory (EMBL) Partnership, Centre for Molecular Medicine Norway, University of Oslo, Oslo, Norway

Artur Cieślar-Pobuda (17,37), Nordic European Molecular Laboratory (EMBL) Partnership, Centre for Molecular Medicine Norway, University of Oslo, Oslo, Norway

Wojciech Gaweł (99), Solveere Sp z o.o., Ogrodzieniec, Poland

Saeid Ghavami (5), Department of Human Anatomy and Cell Science, Max Rady College of Medicine, University of Manitoba, Winnipeg, MB, Canada; Health Policy Research Center, Institute of Health, Shiraz University of Medical Sciences, Shiraz, Iran

Joanna Gola (143), Department of Molecular Biology, School of Pharmacy with the Division of Laboratory Medicine in Sosnowiec, Medical University of Silesia, Katowice, Poland

Catherine Grillon (203), Center for Molecular Biophysics, UPR4301 CNRS CS80054, Orleans Cedex 2, France

Andrzej Hudecki (85,110), Institute of Non-Ferrous Metals, Gliwice, Poland

Jolanta Hybiak (73), Department of Pathomorphology, Pomeranian Medical University in Szczecin, Szczecin, Poland

Claudine Kieda (73), Department of Oncology with Laboratory of Molecular Oncology, Military Institute of Medicine, Warsaw, Poland

Gerard Kiryczyński (85), Department of Pathomorphology, Pomeranian Medical University in Szczecin, Szczecin, Poland

Laura D. Los (1), LinkoCare Life Sciences AB (Ltd), Linköping, Sweden

Marek J. Łos (5,85,191,203), Małopolska Centre of Biotechnology, Jagiellonian University, Krakow, Poland; Biotechnology Centre, Silesian University of Technology, Gliwice, Poland; Department of Pathology, Pomeranian Medical University, Szczecin, Poland; LinkoCare Life Sciences AB (Ltd), Linköping, Sweden; Center for Molecular Biophysics, UPR4301 CNRS CS80054, Orleans Cedex 2, France

Batoryna Olgierd (163), Department of Community Pharmacy, Faculty of Pharmacy with the Division of Laboratory Medicine in Sosnowiec, Medical University of Silesia, Katowice, Poland

Soumya Panigrahi (203), Case Western Reserve University, Cleveland, OH, United States

Mehrdad Rafat (191), Department of Biomedical Engineering, Linköping University; LinkoCare Life Sciences AB (Ltd), Linköping, Sweden

Frank Schweizer (99), Department of Chemistry, University of Manitoba, Winnipeg, MB, Canada

Katarzyna Sielatycka (203), Department of Physiology, Faculty of Biology, University of Szczecin, Szczecin, Poland

Bartosz Sikora (191), Department of Molecular Biology, School of Pharmacy with the Division of Laboratory Medicine in Sosnowiec; Department of Cytophysiology, Chair of Histology and Embryology, School of Medicine in Katowice, Medical University of Silesia, Katowice, Poland

Paulina Siwek (163), Department of Community Pharmacy, Faculty of Pharmacy with the Division of Laboratory Medicine in Sosnowiec, Medical University of Silesia, Katowice, Poland

Aleksandra Sklarek (163), Department of Community Pharmacy, Faculty of Pharmacy with the Division of Laboratory Medicine in Sosnowiec, Medical University of Silesia, Katowice, Poland

Aleksandra Skubis (5), Department of Cytophysiology, Chair of Histology and Embryology, School of Medicine in Katowice; Department of Molecular Biology, School of Pharmacy with the Division of Laboratory Medicine in Sosnowiec, Medical University of Silesia, Katowice, Poland

Judith Staerk (17,37), Nordic European Molecular Laboratory (EMBL) Partnership, Centre for Molecular Medicine Norway, University of Oslo; Department of Haematology; Norwegian Center for Stem Cell Research, Department of Immunology, Oslo University Hospital, Oslo, Norway

Ewa Waluga (163), Department of Community Pharmacy, Faculty of Pharmacy with the Division of Laboratory Medicine in Sosnowiec, Medical University of Silesia, Katowice, Poland

Emilia Wiecheć (1,73), Department of Clinical and Experimental Medicine (IKE), Division of Cell Biology, Linköping University, Linköping, Sweden

Magdalena Wierzbik-Strońska (110), Katowice School of Technology, The University of Science and Art, Katowice, Poland

Renata Wilk (121), Department of Anatomy, School of Health Sciences in Katowice, Medical University of Silesia, Katowice, Poland

Acknowledgments

The editors MJŁ and AH acknowledge the support from NCN grant #: 2016/21/B/NZ1/02812 and NCBIR #POIR.04.01.02-00-0085/17. HD also acknowledges the support from the internal grant 25/G/S/2016, MJŁ acknowledges the support from LE STUDIUM Institute for Advanced Studies (region Centre-Val de Loire, France) through its Smart Loire Valley General Program, cofunded by the Marie Skłodowska-Curie Actions, grant # 665790. EW acknowledges the support from The Swedish Cancer Society (2017/301) and the County Council of Östergötland.

Introduction

Regenerative medicine is a relatively new field of study that treats injuries and diseases by harnessing the body's own regenerative capabilities. Development of regenerative medicine is mainly motivated by diseases, injuries, and loss of organs; therefore, the aim of regenerative medicine is to mitigate the shortage of available allo-, and autografts, so that diseased and damaged tissues and whole organs could potentially be regenerated or replaced by a functioning substitute generated in vitro.

Encompassing a broad range of scientific disciplines (from molecular biology and genetics to immunology and biochemistry), regenerative medicine research also includes such topics as the physiological mechanisms of the regulation of growth of stem cells and what makes stem cells change into other cell types (differentiation processes). Relatively recently, also the transdifferentiation procedures (change of one cell type into another while completely bypassing the stem cell stage in most cases) are increasingly gaining attention, as they offer additional safety and are typically less time consuming than traditional differentiation techniques utilizing iPS cells as a progeny.

There are three main ways regenerative medicine could be applied to patients: ex vivo organ growth, self-repair, and cellular therapies.

Combining stem cells with biomaterial scaffolds provides a promising strategy for engineering tissues and cellular delivery. Biomaterials development is rapidly advancing to display properties that, in a precise and physiological fashion, could drive stem-cell fate both in vitro and in vivo.

DESCRIPTION

Stem Cells and Biomaterials for Regenerative Medicine addresses the urgent need for a compact source of information on both the cellular and biomaterial aspects of regenerative medicine. By developing a mutual understanding between three separately functioning areas of science—medicine, the latest technology, and clinical economics—the volume encourages interdisciplinary relationships that will lead to solutions for the significant challenges faced by today's regenerative medicine. Users will find sections on the homeostatic balance created by apoptosis and proliferating tissue stem cells, the naturally regenerative capacities of various tissue types, the potential regenerative benefits of iPS generation, various differentiation protocols, and more.

Written in easily accessible language, this volume is appropriate for any professional or medical staff looking to expand their knowledge with regard to stem cells and regenerative medicine, as well as a non-professional with some background knowledge, interested in gaining additional more in-depth information in the areas related to regenerative medicine.

KEY FEATURES

- Provides readers with key information on tissue engineering, artificial organs, and biomaterials, while using broadly accessible language.
- Provides broad introduction to, and examples of, various types of stem cells, core concepts of regenerative medicine, biomaterials, nanotechnology and nanomaterials, somatic cell transdifferentiation, and more.
- Edited and authored by researchers with expertise in regenerative medicine, (cancer) stem cells, biomaterials, genetics, and nanomaterials.

READERSHIP

Researchers and students in stem cell biology, regenerative medicine, tissue engineering, and biomaterials; general public, with advanced interest in the topics covered by the book.

List of Biopolymers Discussed in the Book

Biopolymer	Shortcut	Chapters
Alumina	Al_2O_3	7, 8
Calcium phosphate		7
Cobalt-chromium alloys		7
Chitosan		13
Glass ceramics		7
Hydroxyapatite	HA	5
Nylon 6/6		7
Polyamide	PA	8
Poly(ethylene terephthalate)		7
Polylactide	PLA	13
Poly(lactic acid)		7
Polycaprolactone	PCL	8, 13
Poly-DL-lactic-*co*-glycolic acid	PLGA	5, 13
Poly(ethylene glycol) dimethacrylate	PEGDMA	5
Poly(ethylene oxide)	PEO	7
Polyglycolic acid	PGA	5, 13
Poly(methyl methacrylate)	PMMA	7
Polypropylene		7
Polytetrafluoroethylene		7
Polyvinyl alcohol	PVA	7
Pyrolytic carbon		7
Silicon monoxide	SiO_2	8
Silicon rubber		7

Biopolymer	Shortcut	Chapters
Stainless steels		7
Titanium alloys		7
Tricalcium phosphate	TCP	5
Ultra-high-molecular-weight polyethylene	UHMWPE	7
Zirconia	ZrO_2	7

Chapter 1

Introduction and Historic Perspective

Karolina Bakalorz[*], Laura D. Los[†], Emilia Wiecheć[‡]
*Faculty of Chemistry, Silesian University of Technology, Gliwice, Poland
†LinkoCare Life Sciences AB (Ltd), Linköping, Sweden
‡Department of Clinical and Experimental Medicine (IKE), Division of Cell Biology, Linköping University, Linköping, Sweden

Regenerative medicine is a relatively young field of multidisciplinary research that involves replacing, repairing, or regeneration of impaired body organs, tissues, and cells. The first mentions about regenerative medicine come most likely from the paper "The future of multihospital systems" written by a recognized futurist and acknowledged authority on the changing American healthcare system, namely, Dr. Leland Kaiser. Already in 1992, he paid attention to a new branch of medicine that would influence the cure for chronic diseases and "regenerate tired and failing organ systems" [1]. The fascination of Prometheus's liver that was able to regenerate in a short time after being eaten by an eagle or the regrowth of salamandra's (*Ambystoma mexicanum*) amputated limb has become an inspiration for the current regenerative medicine [2]. In the case of a serious trauma or disease, transplantation of the missing or impaired organ is the only salvation for the patients. However, due to the shortage of organ donations and severe side effects of lifelong immunosuppression, the regenerative approach is currently gaining importance [3, 4]. It is likely that regenerative medicine will complement and then replace traditional transplantology in the near future.

There are several approaches, which are endeavoring to aid the objective of regenerative medicine. Nowadays, the progress of rapidly evolving regenerative medicine field is mainly associated with stem-cell-based therapy and tissue engineering [5]. Stem cells are undifferentiated cells, which possess unlimited capability to divide, self-renew, and differentiate into other types of cells within the body. The differentiation potential of stem cells during the in vivo maturation of the zygote goes down; therefore, the blastocyst-derived pluripotent embryonic stem cells have a huge regenerative potential and are a subject of extensive research [6, 7]. However due to their implication in carcinogenesis (teratoma formation), immune intolerance (external source; allogenic) as well as ethical

Stem Cells and Biomaterials for Regenerative Medicine. https://doi.org/10.1016/B978-0-12-812258-7.00001-0

issues (embryo-derived), the new strategies to obtain autologous source of stem cells are in progress. Several years of research on cloning of mammals and embryonic stem cells have allowed the identification of genes, which are responsible for the pluripotency of cells. In 2006, the breakthrough discovery in stem cell research was presented by Shinya Yamanaka from the Kyoto University and had been awarded the Nobel prize in physiology or medicine in 2012. He has shown that genetically modified somatic cells can be reprogrammed into induced pluripotent stem cells (iPSCs) and they closely resemble embryonic stem cells, which might overcome the controversy with destruction of embryos [8]. Nevertheless, the major barrier facing clinical applications of iPSCs, besides their implication in teratoma formation, is use of viral vectors and recombinant DNA in order to express the pluripotent-cell-specific transcription factors during reprogramming. The new technologies are evolving and include among others the use of recombinant-cell-penetrating peptides and direct transfection of synthetic mRNA, which encode appropriate transcription factors [9, 10]. Furthermore, successful attempts to create various types of differentiated cells from fibroblasts without passing through the stem cell state (transdifferentiation) have been performed [11–14].

Unfortunately, only a small population of transplanted stem cells usually survive in the given microenvironment, and only those cells are capable of further differentiation. The classical injections of cellular suspension in saline vehicles are not an optimal therapeutic approach and in most of the cases <10% of transplanted cells is retained in the target site [15]. The important issue is to create a unique microenvironment capable of supporting implanted cells in signaling molecules, growth factors, and hormones. A good carrier of such substances is biomaterial-based scaffold, which additionally provides a mechanical support for the cells [16–18]. Tissue engineering is a branch of regenerative medicine and the advances in the development of biomaterials and biomaterial scaffolds in combination with stem cells are capable of promoting regeneration [5, 18].

Regenerative medicine and stem cells give a great hope in the treatment of currently incurable diseases. During the last years, a number of new strategies to efficiently obtain and maintain certain types of stem cells for various treatments such as neurodegenerative diseases, heart disease and osteoporosis have been described [19–21]. However, the yield of in vitro stem cell production and their propensity to induce tumors upon transplantation is a serious challenge for the researchers. We can still benefit from the current knowledge in terms of using stem cells for disease modeling and drug screening on the way to personalized medicine [22, 23].

Despite so many efforts, there are not many stem-cell-based therapies approved that are commercially available [18, 24]. Since the pioneer bone marrow transplant between a related donor and recipient, which was performed in 1956 by Dr. E. Donnall Thomas in New York, this is still the most widely used stem-cell therapy nowadays. Moreover, the attempts to transplantation of artificial tissues and organs on routine clinical settings are rapidly developing. The best described examples involve biomaterial-based bladder reconstruction, vascular regeneration, skin substitutes, tissue-engineered trachea, and artificial corneas [25–33].

REFERENCES

[1] Kaiser LR. The future of multihospital systems. Top Health Care Financ 1992;18(4):32–45.
[2] Kragl M, et al. Cells keep a memory of their tissue origin during axolotl limb regeneration. Nature 2009;460(7251):60–5.
[3] Saidi RF, Hejazii Kenari SK. Clinical transplantation and tolerance: are we there yet? Int J Organ Transplant Med 2014;5(4):137–45.
[4] Power C, Rasko JE. Whither prometheus' liver? Greek myth and the science of regeneration. Ann Intern Med 2008;149(6):421–6.
[5] Katari R, Peloso A, Orlando G. Tissue engineering and regenerative medicine: semantic considerations for an evolving paradigm. Front Bioeng Biotechnol 2014;2:57.
[6] Mahla RS. Stem cells applications in regenerative medicine and disease therapeutics. Int J Cell Biol 2016;2016:6940283.
[7] Passier R. Potential of human embryonic stem cells in regenerative medicine. Horm Res 2003;60(Suppl. 3):11–4.
[8] Takahashi K, Yamanaka S. Induction of pluripotent stem cells from mouse embryonic and adult fibroblast cultures by defined factors. Cell 2006;126(4):663–76.
[9] Heng BC, Fussenegger M. Integration-free reprogramming of human somatic cells to induced pluripotent stem cells (iPSCs) without viral vectors, recombinant DNA, and genetic modification. Methods Mol Biol 2014;1151:75–94.
[10] Rony IK, et al. Inducing pluripotency in vitro: recent advances and highlights in induced pluripotent stem cells generation and pluripotency reprogramming. Cell Prolif 2015;48(2):140–56.
[11] Xu Z, et al. Direct conversion of human fibroblasts to induced serotonergic neurons. Mol Psychiatry 2016;21(1):62–70.
[12] Son EY, et al. Conversion of mouse and human fibroblasts into functional spinal motor neurons. Cell Stem Cell 2011;9(3):205–18.
[13] Xu A, Cheng L. Chemical transdifferentiation: closer to regenerative medicine. Front Med 2016;10(2):152–65.
[14] Guo J, Wang H, Hu X. Reprogramming and transdifferentiation shift the landscape of regenerative medicine. DNA Cell Biol 2013;32(10):565–72.
[15] Mooney DJ, Vandenburgh H. Cell delivery mechanisms for tissue repair. Cell Stem Cell 2008;2(3):205–13.
[16] Edgar L, et al. Heterogeneity of scaffold biomaterials in tissue engineering. Materials (Basel) 2016;9(5). pii:E332.
[17] Jammalamadaka U, Tappa K. Recent advances in biomaterials for 3D printing and tissue engineering. J Funct Biomater 2018;9(1). pii: E22.
[18] Mao AS, Mooney DJ. Regenerative medicine: current therapies and future directions. Proc Natl Acad Sci USA 2015;112(47):14452–9.
[19] Lunn JS, et al. Stem cell technology for neurodegenerative diseases. Ann Neurol 2011;70(3):353–61.
[20] Tompkins BA, et al. Preclinical studies of stem cell therapy for heart disease. Circ Res 2018;122(7):1006–20.

[21] Phetfong J, et al. Osteoporosis: the current status of mesenchymal stem cell-based therapy. Cell Mol Biol Lett 2016;21:12.
[22] Avior Y, Sagi I, Benvenisty N. Pluripotent stem cells in disease modelling and drug discovery. Nat Rev Mol Cell Biol 2016;17(3):170–82.
[23] Suh W. A new era of disease modeling and drug discovery using induced pluripotent stem cells. Arch Pharm Res 2017;40(1):1–12.
[24] Reisman M, Adams KT. Stem cell therapy: a look at current research, regulations, and remaining hurdles. P&T 2014;39(12):846–57.
[25] Atala A, et al. Tissue-engineered autologous bladders for patients needing cystoplasty. Lancet 2006;367(9518):1241–6.
[26] Lam Van Ba O, et al. Bladder tissue engineering: a literature review. Adv Drug Deliv Rev 2015;82–83:31–7.
[27] Li S, Sengupta D, Chien S. Vascular tissue engineering: from in vitro to in situ. Wiley Interdiscip Rev Syst Biol Med 2014;6(1):61–76.
[28] Shin'oka T, Imai Y, Ikada Y. Transplantation of a tissue-engineered pulmonary artery. N Engl J Med 2001;344(7):532–3.
[29] Kurobe H, et al. Concise review: tissue-engineered vascular grafts for cardiac surgery: past, present, and future. Stem Cells Transl Med 2012;1(7):566–71.
[30] Vig K, et al. Advances in skin regeneration using tissue engineering. Int J Mol Sci 2017;18(4). pii:E789.
[31] Elliott MJ, et al. Stem-cell-based, tissue engineered tracheal replacement in a child: a 2-year follow-up study. Lancet 2012;380(9846):994–1000.
[32] O'Leary C, et al. Respiratory tissue engineering: current status and opportunities for the future. Tissue Eng Part B Rev 2015;21(4):323–44.
[33] Ghezzi CE, Rnjak-Kovacina J, Kaplan DL. Corneal tissue engineering: recent advances and future perspectives. Tissue Eng Part B Rev 2015;21(3):278–87.

Chapter 2

Stem Cells

Marek J. Łos*,†,‡, Aleksandra Skubis#,§, Saeid Ghavami¶,||

*Małopolska Centre of Biotechnology, Jagiellonian University, Krakow, Poland

†LinkoCare Life Sciences AB (Ltd), Linköping, Sweden

‡Center for Molecular Biophysics, UPR4301 CNRS CS80054, Orleans Cedex 2, France

§Department of Molecular Biology, School of Pharmacy with the Division of Laboratory Medicine in Sosnowiec, Medical University of Silesia, Katowice, Poland

¶Department of Human Anatomy and Cell Science, Max Rady College of Medicine, University of Manitoba, Winnipeg, MB, Canada

||Health Policy Research Center, Institute of Health, Shiraz University of Medical Sciences, Shiraz, Iran

#Department of Cytophysiology, Chair of Histology and Embryology, School of Medicine in Katowice, Medical University of Silesia, Katowice, Poland

INTRODUCTION

Stem cells are undifferentiated cells defined by two criteria: a potential to self-renewal and differentiate into multiple mature cell types. Self-renewal is reflected by their capacity to undergo multiple/limitless divisions [1, 2]. Several signaling pathways are involved in self-renewal of stem cells, that is, Notch, Wnt, and Hedgehog pathways or Polycomb family proteins [2]. Stem cells may replicate through two types of cell divisions: symmetric: (i) both daughter cells are identical as parent stem cells, and (ii) asymmetric, one of two cells is identical to parent stem cells and the second enters differentiation into more specialized cell lineage (Fig. 2.1).

Symmetric divisions assure repopulation of the niche of undifferentiated stem cells. Asymmetric divisions warrant preservation of the stem cell pool and maintain homeostasis within the given tissue [3, 4]. Differentiation of a subpopulation of stem cell descendants is crucial for the regeneration and repair of damaged tissues and organs in the body [2, 5].

Stem cell niches are located in specific anatomic microenvironment that allows for stem cells self-renewal and maintenance of their undifferentiated potential [6–8]. Stem cell niches are characterized by defined histologic environment, the availability of certain types of adhesion molecules, and often also the presence of supporting cells [8]. Each stem cell niche has organ-specific histologic features. As examples may serve, niche of hematopoietic stem cells (HSC) in trabecular bones, intestinal crypts in digestive tract, hair follicle bulge

Stem Cells and Biomaterials for Regenerative Medicine. https://doi.org/10.1016/B978-0-12-812258-7.00002-2

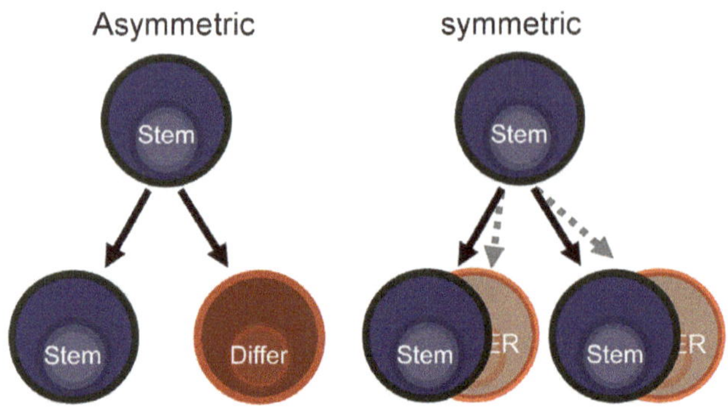

FIG. 2.1 Asymmetric and symmetric division of stem cells.

for regeneration of keratinocytes, hair, and sebaceous glands or subventricular zone of the cerebellum, and the subgranular zone of the hypothalamus is responsible for producing astrocytes [9–11].

Functionally, stem cell niches may be classified as either simple niche, complex niche, or storage niche (Fig. 2.2) [11]. In simple niche, stem cells adhere to

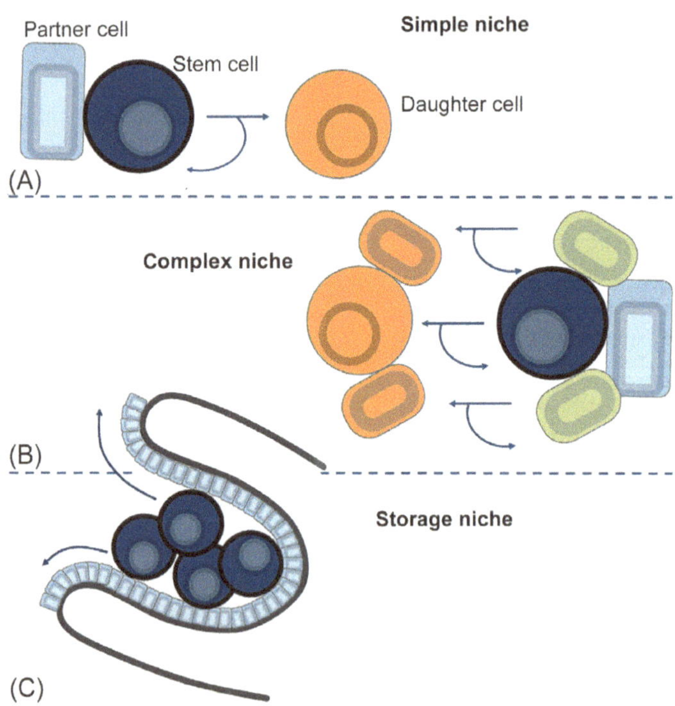

FIG. 2.2 Niche types: simple (A), complex (B), and storage (C). *(Based on Ohlstein B, Kai T, Decotto E, Spradling A. The stem cell niche: theme and variations. Curr Opin Cell Biol 2004;16(6):693–9.)*

each other through specific adhesion molecules, which lock stem cell in a place until proliferation signal is received. They are simple in structure and are typical for epithelial, digestive tissues, and gonads. Complex niches consist of stem cell interacting with other stem cells and differentiated partner cell, while forming large complexes of cells, which provide better regulatory control mechanisms than in a simple niche, i.e., subventricular zone neural stem cells reside in a complex niche. Storage niches are basically repositories of stem cells. There, the cells typically do not divide, instead they migrate from the niche after receiving signals specific signals. Hair follicle and niches storing melanoblast progenitors niche are examples of storage niche [11].

Stem Cell—Potency

Stem cells could be classified according to their plasticity/regenerative potential into: totipotent, pluripotent, multipotent, oligopotent, and unipotent (Table 2.1) [5, 12].

Totipotent cells have ability to differentiate into cell linages from all three germ layers: mesoderm, endoderm, and ectoderm including placental cell, and this capability has only zygote and first blastomere in human organism. During divisions, morula and then blastocyst is formed. Cells composing blastocyst's inner cell mass (ICM) have pluripotent properties. Induced pluripotent stem cells (iPS) have ability to differentiate into cells from all three germ layers, with the exception of extrafetal tissues (i.e., placenta). In human organism, only embryonic stem cells, which form ICM, have pluripotent ability. Multipotent stem cells differentiate into many types of cells that originate from the same germ layer (Fig. 2.3) [5, 12].

Types of Stem Cell

Stem cells could be classified also according to their origin. We distinguish embryonic stem cells, which are isolated only at stage of embryonic development,

TABLE 2.1 Classification of Stem Cells

Cell Potency	Characteristic
Totipotent	Ability to differentiate into cell linages from all three germ layers: mesoderm, endoderm, and ectoderm including placental cell
Pluripotent	Ability to differentiate into cells from all three germ layers with the exception of placenta
Multipotent	Ability to differentiate into a limited number of types from one germ layer
Oligopotent	Ability to differentiate into a few type of cells
Unipotent	Ability to produce cells of their own type

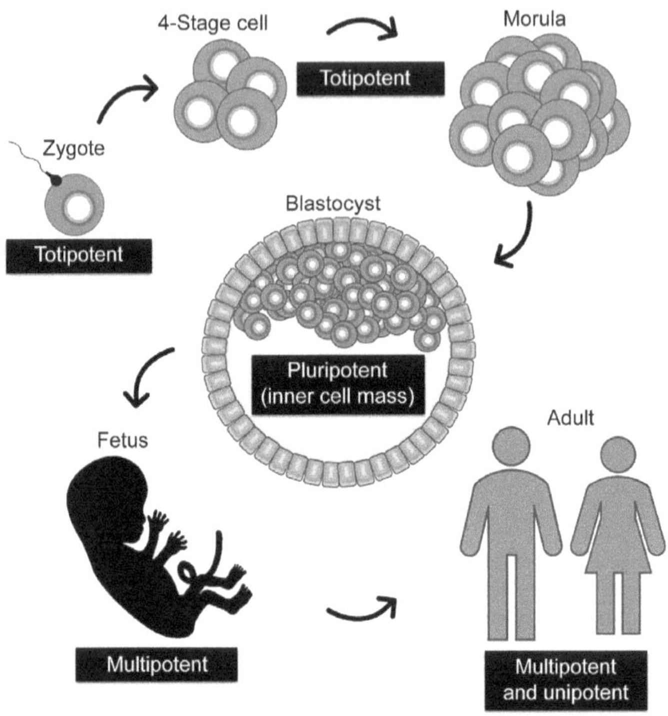

FIG. 2.3 Stem cell potency.

and adult stem cells, which are located in organs of mature body. Another type of stem cells are called iPS. They are generated from somatic cells and have the properties similar to embryonic stem cells. Types of stem cells with their sources indicated, characteristic properties, and the most important identification markers are shown in Table 2.2.

Embryonic Stem Cells

Embryonic stem cells (ESC) are pluripotent cells isolated from ICM of late blastocysts. They have capacity to differentiate into cell from all three germ layers. They were first isolated from mice embryos in 1981. However, human ES were first obtained just in 1998. ESC's use is limited mainly due to the ethical concerns, since most methods of their isolation require the destruction of embryo [1, 13–15]. Culture of ESC is difficult and requires feeder cells, for example murine embryonic fibroblasts, to maintain them in undifferentiated state [4, 13, 16]. Moreover, self-renewal and multilineage differentiation potential often deteriorates during long-term cell culture. Their use in regenerative therapies is also limited by two factors. Main problem is still identification of ESC in ESC-derived tissues, since ESC contamination of such tissues carries the risk of teratomagenesis. It is also hard to distinguish ESC from other types of stem

TABLE 2.2 Types of Stem Cells

	Source	Characteristic	Markers
Embryonic stem cells (ESC)	Blastocysts	• Pluripotent • Endless proliferation • Use ethically questionable in many countries • Due to the lack of complete immunocompatibility, organs and tissues generated from them, will likely be immunorejected • Risk of teratoma formation upon injection in undifferentiated state	CD133, CD31, CD59, SSEA-1, SSEA-3,SSEA-4, Oct-3/4, Sox2, KLF-4, Nanog, TRA-1-60, TRA-1-81
Adult stem cells (ASC)	Adipose tissue/bone marrow/peripheral blood	• Multipotent • No risk of rejection during autotransplantation • Less/no risk of tumor formation • Limited number in tissue	*Examples* HSC: c-Kit, CD34, Sca-1 Thy-1, CD133 MSC: CD90, CD105, CD73, CD44, CD117, STRO-1
Induced pluripotent stem cells (iPS)	Virtually any type of cells, most commonly: fibroblasts, keratinocytes, peripheral blood mononuclear cells	• Pluripotent • Obtained through reprogramming of somatic cell • Low rate of reprogramming • No ethical problems • Personal regenerative medicine • Low risk of immune rejection • Risk of teratoma formation upon injection in undifferentiated state	Oct3/4, Nanog, Dax1, Ras, Zfp296

HSC, hematopoietic stem cells; *MSC*, mesenchymal stem cells.

cells and cancer cells. Recently discovered differential expression of PFKFB3 in cancer stem cells, and iPS, raises hope for the development of tests that would facilitate easy way to distinguish between those cell types [17]. ESC markers are classified into four basic groups: cell surface markers, transcription factors, signal pathway-related intracellular markers, like PFKFB3, and peptides markers. Unfortunately, embryonic stem cell markers also change under cell culture conditions [14].

Cell surface markers are the most important for practical purposes, because of easy possibility of staining and hence, identification ESC, by flow cytometry, and for live cell sorting. The most commonly used cell surface markers are: stage specific embryonic antigens (SSEA-1, -3, -4), cluster of differentiation antigens (CD133, CD31, CD59, CD49f), TRA-1-60 and TRA-1-81; secreted: stem cell factor (SCF), and teratocarcinoma-derived growth factor-1. Among transcription factors, we distinguish genes, which are involved in regulation of important stemness pathways, for example, Oct-3/4, Sox2, KLF-4, and Nanog. Signaling-related intracellular markers play relevant role in maintaining ESC self-renewal. As examples may serve proteins of Wnt/β-catenin signaling pathway, SMAD proteins, and molecules associated to transforming growth factor-β (TGF-β) [14].

To ensure a safe use of ESC and iPS for therapeutic purposes, it is important to periodically check their genetic stability and pluripotency in vitro. Genetic stability is essential but may be impaired through long-term culture in vitro. Standard methods to assessment of genetic stability are: karyotyping, sequencing of whole genome, and whole-genome genotyping analysis. All these methods have some advantages and limitations, for example, high costs, the size of abnormalities in genome that a given method allows for detection; therefore, the use of two methods simultaneously is recommended [13].

Testing ESC or iPS for pluripotency has been traditionally done by teratoma assays. Typically 1 million of ESC is injected into immunodeficient mice and then forming of tumors is observed after around 6 weeks. Tumor tissue should represent histologically elements of mesoderm, ectoderm, and endoderm tissues to confirm pluripotency of the tested cells [13].

ESC are essential in research of human development, the process of differentiation at early stage of embryonal life, and for the development of models of genetic disorders [4]. It was demonstrated recently that ESC could be used in regeneration of cardiomyocytes, neuronal cells, and pancreas islet cells [15]. However, their use in clinical trials is impeded due to often heterogeneous population of cells and risk of formation a teratoma or immune rejection [16].

Adult Stem Cells

Adult stem cells, also called tissue-specific stem cells or somatic stem cells, are multipotent cells, which are responsible for replacing damaged and dead cells in the body (tissue maintenance role) [18]. They are found in specific microenvironment called stem cell niche. Adult stem cells are often isolated from bone marrow, blood, adipose tissue, liver and skin [5, 16]. However, as a natural consequence of aging, their quantity and quality decreases with the age. Aging of stem cells affect regenerative potential, growth, and divisions [19].

MESENCHYMAL STEM CELLS

Mesenchymal stem cells (MSC) are nonhematopoietic, multipotent cells, isolated from different sources: bone marrow, adipose tissue, skin, peripheral blood but also from perinatal tissue as: umbilical cord blood, amniotic fluid, membrane, and placenta [5, 9, 12, 20, 21].

The International Society for Cellular Therapy and the International Federation for Adipose Therapeutics created minimal criteria to define MSC. They must be fibroblast-like and adhere to cell-culture plastic vessels. They have to differentiate into osteocytes, chondrocytes, and adipocytes. Differentiation must be demonstrated by staining Alizarin Red for osteoblasts, Oil Red O for adipocytes, and Alcian blue for chondrocytes [5, 12, 22, 23]. Moreover, human MSC are identified based on cell surface markers. They must express CD73 (ecto 5′ nucleotidase), CD90 (Thy-1), and CD105 (endoglin) in over 95% of the MSC population and they must lack of expression in <2% of cell population of CD45 (leukocyte common antigen), CD34 (hematopoietic progenitor cell antigen), CD14 and CD11b (expressed on monocytes and macrophages), CD79a or CD19 (markers of B-cells), and HLA class II proved by flow cytometry analysis [23, 24]. There are also characteristic markers other than those mentioned: Stro-1, cell membrane single pass type I protein, CD271 (low-affinity nerve growth factor receptor), CD146 (melanoma cell adhesion molecule), and SSEA-4 (cell surface glycosphingolipids). However, they are not universal markers because their expression depends on source of MSC [24]. MSC derived from various sources have different profile of cell surface and differentiation potential [9, 12, 24, 25]. For example, bone marrow MSC have the strongest potential to differentiate into chondrocytes and osteoblasts, and adipose derived MSC into adipocytes [9, 25].

Since adipose-derived stem cells (ADSC) have multipotent properties and adipose tissue donation is less invasive than bone marrow biopsy and the procedure allows for harvesting more cells they are a very promising source of cellular precursors for the use in regenerative medicine. Furthermore, ADSC does not pose ethical problems as they are often a medical waste and/or a by-product of cosmetic operations. They could potentially be used for medical procedures in bones cartilage myocardial and skin tissue regeneration [22, 26–28].

Bone-marrow-derived stem cells are obtained mostly from bone marrow aspiration from iliac bone crest, or sternum. The extraction procedure is however quite painful and abundance of stem cells is about 0.01% of bone marrow cells, so unless absolutely necessary, other sources of MSC should be used [26, 27].

MSC could potentially be used for the therapeutic interventions in many diseases. Nowadays MSC are used in the clinic for mainly in bone reconstruction therapy. For this purpose, autologous MSC, usually from bone marrow, are used [9, 21]. MSC transplantation has also been tested to repair damaged cartilage. In such attempts, MSC often enriched with biocompatible materials, as platelet-rich fibrin glue, hyaluronic acid and collagen membrane are applied [9, 21]. Several attempts have been made to use MSC in cardiovascular injury treatment, like for example myocardial infarction or chronic ischemic cardiomyopathy [21].

MSC have also immunomodulatory functions through secreting various factors: cytokines, chemokines, and growth factors [9]. MSC act on immunological system through paracrine activity and release molecules like: IL-1, IL-2, IL-12 (interleukin), TNF-α (tumor necrosis factor-α), and INF-γ (interferon-gamma). They act mostly antiinflammatory through regulation of T-cells, B-cells, NK cells, monocytes, and macrophages [21].

MSC secrete also molecules, which promote local angiogenesis, proliferation and inhibit oxidative stress and apoptosis. Antiapoptotic activity of MSC is extremely important for stem cells after exposure to chemical and physical damage, for example, drugs and other chemicals, radiation, or mechanical injury [21, 25]. MSC have also a natural ability to migrate into injury site [9].

HEMATOPOIETIC STEM CELLS

HSC are multipotent cells, which can differentiate into progenitor and mature blood cells. They are responsible for hematopoiesis and produce immune cells in organism. HSC are defined based on specific markers: c-Kit, CD34, Sca-1 Thy-1, CD133 [29]. Like MSC, they have unique ability to migrate into damaged tissues [30]. Transplantation of HSC is a treatment of choice particularly aggressive leukemias [31]. For harvesting purposes, they are "forced" to be release from bone marrow by cytokines as granulocyte-colony-stimulating factor, thrombopoietin or c-Kit ligand, and collect from blood by leukapheresis [30, 32]. Their stem cell niche is located in bone marrow; in particular, in the arteriolar niche and the sinusoidal-megakaryocyte niche [19, 32, 33]. Niche of HSC is rich in various types of cells such as pericytes, endothelial cells, osteolineage cells, sympathetic nerves, and Schwann cells, which help to maintain the undifferentiated state of HSC [8]. HSC niche is regulated through different factors, for example, SCF, TGF-β, and CXC motif chemokine 12 also called stromal cell-derived factor 1 [32].

OTHER TYPES OF ADULT STEM CELLS

Liver is an organ with enormous regenerative capacity. Regeneration of liver is attributed mainly to the high proliferation capacity of mature hepatocytes. However, hepatic tissue is also rich in *hematopoietic stem cells* (*HSC*), which are responsible for regrowth of liver when hepatocytes are damaged through severe disease. These cells can differentiate into biliary epithelial cells and also hepatocytes, so they are bipotent. Stem cell niches are located at the periphery of bile ducts. They express surface markers such as cytokeratins-8, -18, and -19, epithelial cell adhesion molecule, neural cell adhesion molecule (NCAM, also called CD56) and CD133 [34, 35].

Skeletal stem cells (*satellite cells, SSC*) are located in bone marrow, and local periosteum, and they are responsible for regeneration of bones, adipocytes, and cartilages. They have multipotent potential and self-renewal ability. SSC might be used in promoting angiogenesis because they contribute to the stability of blood vessels. Moreover, they can be used in bone tissue regeneration and as a therapeutic component of HSC in treatment of leukemia or myelodysplastic syndromes [36, 37].

Pancreatic stem cells (*PSC*) are endodermal, tissue-specific cells, which are located in different regions of pancreas, for example, ductal epithelium, exocrine gland tissue, and islet tissue. They are responsible for repair damages and have multidifferentiation potential. Nestin and c-Kit are major markers of PSC [38, 39].

Several other types of tissue-specific adult stem cells exist, which help to maintain tissue repair and homeostasis. Muscle stem cells (satellite cells) are responsible for muscle regeneration [19]. Mentioned earlier, neural stem cells (NSC) are multipotent ectodermal cells located in the neural tissue. NSC differentiate into neurons, astrocytes, and other central nervous system cells, and maintain brain homeostasis [19, 40]. Human cornea stem cells are located in cornea limbus. Limbal stem cells divide and after migration into the surface of cornea, differentiate into adult corneal epithelial cells [41]. They are characterized based on mainly markers like: P63, cytokeratins-3, and -12, and ABCG2 [42].

INDUCED PLURIPOTENT STEM CELLS

The methods of production and the biology of iPS are described in detail in the next chapter. Here, we only briefly outline the topic. iPS are pluripotent cells derived from somatic cells through genetic reprogramming [7, 43–45]. The method of their production was first described in 2006 by Yamanaka and Takahashi, who identified genes, which when transferred into somatic cells, causing a reversion of them into pluripotent state. The four stemness transcription factors, called "Yamanaka Factors" or OSKM factors, are: Oct3/4, Sox2, c-Myc, and Klf4 [46, 47]. For the discovery, Yamanaka was awarded Nobel Prize in *Physiology and Medicine* in 2012. iPS are similar in self-renewal, morphology, growth kinetics, and gene expression to ESC. They also have an ability to differentiate into cells from all three germs layer [1, 15, 43]. The pluripotency of iPSC is confirmed as in ESC with teratomas assays [15].

The most frequently used cells for reprogramming are keratinocytes, fibroblasts, and B- and T-lymphocytes from peripheral blood [48]. Nowadays, a number of methods of reprogramming are available. Several modes of delivery of Yamanaka Factors have been explored: different types of virus (retrovirus, lentivirus, adenovirus, and Sendai virus), synthetic mRNAs (the most efficient but also the most expensive method), miRNA, plasmids and proteins [18, 43, 49, 50]. All nucleic-acid-based methods of Yamanaka Factors delivery (especially retroviruses and lentiviruses) carry the risk of insertional mutagenesis. The risk of tumorigenesis of implanted cells derived from iPS currently limits their clinical applications. The generation and use of iPS does not pose ethical problems as their generation is not associated with the destruction of embryos. Furthermore, tissues engineered from iPS are autologous, as it is relatively easy to generate patients-specific iPS.

iPSC have been tested in regeneration of cardiovascular cells, neural cells as neurons and glial cells in therapy of Parkinson's disease, Alzheimer's disease, amyotrophic lateral sclerosis, Riley-Day syndrome, regeneration of hepatocytes [43], corneal limbal cells, and many other cell types. Cellular reprogramming (iPS production) is an enormous breakthrough in medicine and biology. They have been used not only in regenerative medicine but also in the development of creation disease models in vitro, for drugs screening in pharmacological industry [43, 44], as well as for testing of cosmetic products.

REFERENCES

[1] Menon S, Shailendra S, Renda A, Longaker M, Quarto N. An overview of direct somatic reprogramming: the ins and outs of iPSCs. Int J Mol Sci 2016;17(2):141.

[2] Molofsky AV, Pardal R, Morrison SJ. Diverse mechanisms regulate stem cell self-renewal. Curr Opin Cell Biol 2004;16(6):700–7.

[3] Shahriyari L, Komarova NL. Symmetric vs. asymmetric stem cell divisions: an adaptation against cancer? PLoS One 2013;8(10):e76195.

[4] Martin U, editor. Engineering of stem cells. Berlin, Heidelberg: Springer; 2009.

[5] Kmiecik G, Niklińska W, Kuć P, Pancewicz-Wojtkiewicz J, Fil D, Karwowska A, et al. Fetal membranes as a source of stem cells. Adv Med Sci 2013;58(2):185–95.

[6] Beerman I, Luis TC, Singbrant S, Lo Celso C, Méndez-Ferrer S. The evolving view of the hematopoietic stem cell niche. Exp Hematol 2017;50:22–6.

[7] Cieslar-Pobuda A, Knoflach V, Ringh MV, Stark J, Likus W, Siemianowicz K, et al. Transdifferentiation and reprogramming: overview of the processes, their similarities and differences. Biochim Biophys Acta 2017;1864(7):1359–69.

[8] Farahani E, Patra HK, Jangamreddy JR, Rashedi I, Kawalec M, Rao Pariti RK, et al. Cell adhesion molecules and their relation to (cancer) cell stemness. Carcinogenesis 2014;35(4):747–59.

[9] Shao J, Zhang W, Yang T. Using mesenchymal stem cells as a therapy for bone regeneration and repairing. Biol Res 2015;48(1):62.

[10] Morrison SJ, Scadden DT. The bone marrow niche for haematopoietic stem cells. Nature 2014;505:327–34.

[11] Ohlstein B, Kai T, Decotto E, Spradling A. The stem cell niche: theme and variations. Curr Opin Cell Biol 2004;16(6):693–9.

[12] Maleki M, Ghanbarvand F, Behvarz MR, Ejtemaei M, Ghadirkhomi E. Comparison of mesenchymal stem cell markers in multiple human adult stem cells. Int J Stem Cells 2014;7(2):118–26.

[13] Damdimopoulou P, Rodin S, Stenfelt S, Antonsson L, Tryggvason K, Hovatta O. Human embryonic stem cells. Best Pract Res Clin Obstet Gynaecol 2016;31:2–12.

[14] Zhao W, Ji X, Zhang F, Li L, Ma L. Embryonic stem cell markers. Molecules 2012;17(12):6196–236.
[15] Wu JY. Pluripotent stem cells and skeletal regeneration—promise and potential. Curr Osteoporos Rep 2015;13(5):342–50.
[16] Rippon HJ, Bishop AE. Embryonic stem cells. Cell Prolif 2004;37(1):23–34.
[17] Cieslar-Pobuda A, Jain MV, Kratz G, Rzeszowska-Wolny J, Ghavami S, Wiechec E. The expression pattern of PFKFB3 enzyme distinguishes between induced-pluripotent stem cells and cancer stem cells. Oncotarget 2015;6(30):29753–70.
[18] Seo B, Hong Y, Do J. Cellular reprogramming using protein and cell-penetrating peptides. Int J Mol Sci 2017;18(3):552.
[19] Ahmed ASI, Sheng MHC, Wasnik S, Baylink DJ, Lau K-HW. Effect of aging on stem cells. World J Exp Med 2017;7(1):1.
[20] Fathi E, Farahzadi R. Isolation, culturing, characterization and aging of adipose tissue-derived mesenchymal stem cells: a brief overview. Braz Arch Biol Technol 2016;59:e16150383.
[21] Murphy MB, Moncivais K, Caplan AI. Mesenchymal stem cells: environmentally responsive therapeutics for regenerative medicine. Exp Mol Med 2013;45(11):e54.
[22] Fathi E, Farahzadi R. University of Tabriz I. Isolation, culturing, characterization and aging of adipose tissue-derived mesenchymal stem cells: a brief overview. Braz Arch Biol Technol 2016;59:e16150383.
[23] Dominici M, Le Blanc K, Mueller I, Slaper-Cortenbach I, Marini FC, Krause DS, et al. Minimal criteria for defining multipotent mesenchymal stromal cells: the International Society for Cellular Therapy position statement. Cytotherapy 2006;8(4):315–7.
[24] Lv F-J, Tuan RS, Cheung KMC, Leung VYL. Concise review: the surface markers and identity of human mesenchymal stem cells: markers and identity of MSCs. Stem Cells 2014;32(6):1408–19.
[25] Paulini J, Higuti E, Bastos RMC, Gomes SA, Rangel ÉB. Mesenchymal stem cells as therapeutic candidates for halting the progression of diabetic nephropathy. Stem Cells Int 2016;2016:1–16.
[26] Dai R, Wang Z, Samanipour R, Koo K-I, Kim K. Adipose-derived stem cells for tissue engineering and regenerative medicine applications. Stem Cells Int 2016;2016:1–19.
[27] Joo HJ, Kim J-H, Hong SJ. Adipose tissue-derived stem cells for myocardial regeneration. Korean Circ J 2017;47(2):151.
[28] Klar AS, Zimoch J, Biedermann T. Skin tissue engineering: application of adipose-derived stem cells. Biomed Res Int 2017;2017:1–12.
[29] Moehrle BM, Geiger H. Aging of hematopoietic stem cells: DNA damage and mutations? Exp Hematol 2016;44(10):895–901.
[30] Magnon C, Frenette PS. Hematopoietic stem cell trafficking. In: StemBook. Cambridge, MA: Harvard Stem Cell Institute; 2008.
[31] Frändberg S, Boreström C, Li S, Fogelstrand L, Palmqvist L. Exploring the heterogeneity of the hematopoietic stem and progenitor cell pool in cord blood: simultaneous staining for side population, aldehyde dehydrogenase activity, and CD34 expression: stem cell staining in cord blood. Transfusion 2015;55(6):1283–9.
[32] Calvi LM, Link DC. The hematopoietic stem cell niche in homeostasis and disease. Blood 2015;126(22):2443–51.
[33] Heazlewood SY, Oteiza A, Cao H, Nilsson SK. Analyzing hematopoietic stem cell homing, lodgment, and engraftment to better understand the bone marrow niche: stem cell homing, lodgment, and engraftment. Ann NY Acad Sci 2014;1310(1):119–28.
[34] Zorn AM. Liver development. In: StemBook. Cambridge, MA: Harvard Stem Cell Institute; 2008.

[35] Christ B, Pelz S. Implication of hepatic stem cells in functional liver repopulation. Cytometry 2013;83A(1):90–102.
[36] Bianco P, Robey PG. Skeletal stem cells. Development 2015;142(6):1023–7.
[37] Abou-Khalil R, Yang F, Lieu S, Julien A, Perry J, Pereira C, et al. Role of muscle stem cells during skeletal regeneration: muscle stem cells in bone repair. Stem Cells 2015;33(5):1501–11.
[38] Gong J, Wang Y, Cao Y. Migration path of stem cells involved in the repair of damaged pancreatic tissue caused by pancreatitis. Int J Clin Exp Pathol 2014;7(5):2438–45.
[39] Jiang F-X, Morahan G. Pancreatic stem cells remain unresolved. Stem Cells Dev 2014;23(23): 2803–12.
[40] Shakhova O, Sommer L. Neural crest-derived stem cells. In: StemBook. Cambridge, MA: Harvard Stem Cell Institute; 2008.
[41] Cieslar-Pobuda A, Rafat M, Knoflach V, Skonieczna M, Hudecki A, Malecki A, et al. Human induced pluripotent stem cell differentiation and direct transdifferentiation into corneal epithelial-like cells. Oncotarget 2016;7(27):42314–29.
[42] Huang M, Wang B, Wan P, Liang X, Wang X, Liu Y, et al. Roles of limbal microvascular net and limbal stroma in regulating maintenance of limbal epithelial stem cells. Cell Tissue Res 2015;359(2):547–63.
[43] Hirschi KK, Li S, Roy K. Induced pluripotent stem cells for regenerative medicine. Ann Rev Biomed Eng 2014;16(1):277–94.
[44] Hosoya M, Czysz K. Translational prospects and challenges in human induced pluripotent stem cell research in drug discovery. Cell 2016;5(4):46.
[45] Wasik AM, Grabarek J, Pantovic A, Cieslar-Pobuda A, Asgari HR, Bundgaard-Nielsen C, et al. Reprogramming and carcinogenesis—parallels and distinctions. Int Rev Cell Mol Biol 2014;308:167–203.
[46] Takahashi K, Yamanaka S. Induction of pluripotent stem cells from mouse embryonic and adult fibroblast cultures by defined factors. Cell 2006;126(4):663–76.
[47] Lensch MW, Mummery Christine L. From stealing fire to cellular reprogramming: a scientific history leading to the 2012 Nobel Prize. Stem Cell Reports 2013;1(1):5–17.
[48] Gnecchi M, Stefanello M, Mura M. Induced pluripotent stem cell technology: toward the future of cardiac arrhythmias. Int J Cardiol 2017;237:49–52.
[49] Rao MS, Malik N. Assessing iPSC reprogramming methods for their suitability in translational medicine. J Cell Biochem 2012;113(10):3061–8.
[50] Ebrahimi B. Reprogramming barriers and enhancers: strategies to enhance the efficiency and kinetics of induced pluripotency. Cell Regen (Lond) 2015;4:10.

Chapter 3

Modern Ways of Obtaining Stem Cells

Safak Caglayan*, Theresa D. Ahrens*, Artur Cieślar-Pobuda*, Judith Staerk*,†,‡

**Nordic European Molecular Laboratory (EMBL) Partnership, Centre for Molecular Medicine Norway, University of Oslo, Oslo, Norway*

†Department of Haematology, Oslo University Hospital, Oslo, Norway

‡Norwegian Center for Stem Cell Research, Department of Immunology, Oslo University Hospital, Oslo, Norway

INTRODUCTION

The field of stem cell research is one of the most dynamically developing areas of life science. It becomes more and more evident that animal models hold limitations with respect to modeling of human disease and to develop new treatment options for patients. Embryonic stem (ES) cells and induced pluripotent stem (iPS) cells can renew indefinitely while maintaining their ability to differentiate into all cell types found in the adult body. These properties make pluripotent stem cells an interesting source not only for basic research but also for regenerative medicine. In addition, the ability to derive patient-specific iPS cells and iPS-cell-derived somatic cell types opens an invaluable resource for the development of drug screening and human in vitro model systems in the physiologic and malignant setting. More recently, three dimensional (3D) organoid cultures have been developed for several different tissues. Although these 3D organoid systems are not yet fully functional, such mini-organs may provide a more physiological environment to study developmental processes and to model human disease. In addition, 3D culture systems derived from customized patient-specific iPS cells can serve as an ideal platform to develop personalized drug screens.

This book chapter provides an overview of adult and pluripotent stem cells, and the promise and shortcomings these systems hold for human developmental biology and regenerative medicine.

Stem Cells and Biomaterials for Regenerative Medicine. https://doi.org/10.1016/B978-0-12-812258-7.00003-4

EMBRYONIC STEM CELLS

ES cells are characterized by their ability to propagate unlimited in vitro while remaining pluripotent. Injection of mouse ES cells into host blastocysts leads to the formation of normal and viable interspecific chimeras with contribution to germ cells [1]. This feature allows researchers to use ES cells and generate genetically engineered mice used in biomedical research.

The first ES cell lines from the inner cell mass of the preimplantation mouse embryo were derived in 1981 using batch-selected serum that supported the undifferentiated cell state on inactivated mouse fibroblasts [2, 3]. Later, leukemia inhibitory factor (LIF) was used to supplement feeder cells [4, 5] and inhibition of extracellular signal–regulated kinase (Erk) signaling has been shown to promote ES cell self-renewal [6]. Interestingly, ES cell derivation was initially restricted to certain mouse strains such as SV129; however, recent studies demonstrated that defined culture conditions using Erk and glycogen synthase kinase 3 (Gsk3) inhibitors allow the derivation of ES cells from different mouse strains [7–9].

Due to the easy accessibility, early embryonic development has been extensively studied in the mouse. These studies indicated that segregation of the first lineage cells can be distinguished at the 32-cell morula stage at embryonic day 3.5 (E 3.5) when the extraembryonic trophoectoderm and the inner cell mass (ICM) form distinct cell populations [10]. At E 4.5, the ICM population is then segregated into the early epiblast, which generates the entire fetus, and the primitive endoderm or hypoblast, which forms the extraembryonic endoderm, including the yolk sac [11]. Single cells isolated from the E 4.5 epiblast can give rise to pluripotent ES cell lines, which have the capacity to contribute to all cell types of the body including the germline when microinjected into another blastocyst [12, 13]. Hence, the early epiblast represents the developmental ground state with true pluripotency and flexibility to produce all embryonic lineages.

Shortly after implantation into the uterus, the mouse epiblast undergoes a dramatic transformation. It reorganizes into a cup-shaped epithelium, called egg cylinder, which is surrounded by the hypoblast. In female embryos, one of the X chromosomes is randomly inactivated during this stage. Furthermore, epiblast cells become primed for differentiation as a result of their exposure to stimuli from extraembryonic tissues. Therefore, egg cylinder epiblast cells are instructively specified according to their location. These so-called epiblast stem cells represent a pluripotent stem cell type that can be derived from the egg cylinder epiblast embryo in mouse and rat [14, 15]. While epiblast stem cells express pluripotency factors and form teratomas similar to mouse ES cells, they are unable to contribute to chimeric blastocysts with germline transmission. Based on these studies, it was proposed that two phases of pluripotency exist, namely, naive and primed, representing epiblast progenitors in the blastocyst and pluripotent progenitors of the gastrulating embryo, respectively [12, 16].

Human ES cells are established from the preimplantation/periimplantation stage embryo [17]. Their initial characterization showed that they share similar features to mouse ES cells such as expression of pluripotency factors and formation of teratomas when transplanted into immunodeficient mouse [17]. However, further studies revealed significant morphological and molecular differences between human and mouse ES cells. For example, mouse ES cells form dome-shaped colonies whereas human ES cell colonies are more flattened. Secondly, human ES cells require fibroblast growth factor (FGF) signaling to maintain self-renewal and pluripotency, while mouse ES cells differentiate with FGF and instead require LIF signaling for propagation. Furthermore, unlike mouse ES cells, human ES cells survive poorly as single cells. At the molecular level, the X chromosome status of female human ES cells remained controversial, although certain lines exhibit markers of X chromosome inactivation and monoallelic expression pattern of X-linked genes [18–20].

Designation of human ES cells is based on their blastocyst origin and their ability to develop teratomas. Expression of pluripotency markers and the potential to form teratomas are common features shared between mouse ES and epiblast stem cells, and molecular profiling showed that human ES cells share many features with mouse epiblast stem cells such as dependence on FGF signaling. These studies suggested a postimplantation identity of established human ES cell lines. Since human ES cells are derived from the blastocyst stage embryo, similarities between human ES cells and epiblast stem cells can be explained with that human ES cells are stabilized in primed state due to culture-induced changes during derivation. The primed state of established human ES cell lines might arise from differences in morphogenesis of the postimplantation embryo [21]. Unlike rodents, human embryos do not form egg cylinder following implantation but instead expand as a flattened layer of an embryonic disc. The derivation protocol of stem cell lines includes in vitro outgrowth of human blastocyst explants and these culture conditions might stimulate proceeding of the ICM more readily into the subsequent embryonic disc stage than mouse counterparts proceed into egg cylinder.

Recognition of established human ES cell lines in the primed state led researchers to search for human ES cells at the naive stage. It was hypothesized that naive or ground-state human ES cells might survive better as single cells and thus allow more efficient genetic targeting. In addition, progression of human ES cells from the naive to the primed state can provide a model to study early epiblast development of the human embryo. Indeed, it has recently been shown that the naive human ES cell state can be obtained with transient expression of transgenes or a small molecule cocktail [22, 23]. Resetting ES cells exhibits a rewiring of the transcriptome, epigenome, and metabolic activity toward ground pluripotency that is more similar to ground-state mouse ES cells. Contribution of naive state human ES cells to the preimplantation mouse epiblast has also been also shown [22], although it is not applicable to test whether they can contribute to normal embryonic development. Alternatively, naive state human ES cells can directly be compared to human blastocyst stage embryo. Comparison of transcriptome and epigenome data revealed close similarity between early human embryos and naive ES cells that are generated using a cocktail of small molecules and growth factors [24]. Interestingly, it has been shown that mouse ES cells can be derived and maintained without growth factors and cytokines in a neutralized environment, suggesting that ground state pluripotency might be an intrinsically self-maintained basal cell state [25]. Since resetting human ES cells to the naive state requires extrinsic factors, it remains to be seen whether these cells constitute the true ground state of pluripotency and whether they can be propagated without growth factors in a permissive environment. Nevertheless, the successful derivation of ES cell lines with distinct features proves the existence of different phases of pluripotency in both, mice and humans.

INDUCED PLURIPOTENT STEM CELLS

It is evident that human ES cells hold great promise for research and regenerative medicine. However, ethical issues accompany human ES cell research [26, 27], and there is need to identify approaches to induce pluripotency in somatic cell types. Already in 1962, John Gurdon reported that the epigenome of a differentiated cell can be reversed to a pluripotent state, and that the nucleus of a frog cell implanted into an oocyte can give rise to fully developed animals. In the nineties, somatic cell nuclear transfer (SCNT) was then used to clone the sheep "Dolly" [28], proving the feasibility of cloning in mammals. SCNT has also been used to generate mice from B cell nuclei demonstrating that nuclei from terminally differentiated cells can be reprogrammed [29]. Cell fusion of somatic and ES cells was another approach to demonstrate that cell extracts from pluripotent cells are sufficient to reset the epigenetic memory in somatic nuclei [30]. All these experiments suggested that defined factors are sufficient to induce a pluripotent state. Indeed, a breakthrough by Takahashi and Yamanaka demonstrated that exogenous expression of only four transcription factors, Oct3/4, Sox2, c-Myc, and Klf4 induces pluripotency in mouse embryonic fibroblasts [31]. Only one year later, the Yamanaka lab was able to show that the same combination of transcription factors is sufficient to generate iPS cells from human neonatal fibroblasts [32].

Since their initial discovery, iPS cells have been generated from many different cell types, including fibroblasts, bone marrow cells, peripheral blood, and liver cells (outlined in Table 3.1). iPS cells offer a unique tool enabling the derivation of pluripotent cell lines from healthy donors and patient samples that can be used as model systems to study human development and disease such as neurodegeneration. Numerous iPS cell biobanks from patients diagnosed with Parkinson's disease and other disorders have been created [33, 34], thus providing scientists with a unique resource to study molecular mechanisms underlying Parkinson's pathology and to assess different types of therapies. For example, transplantation of dopaminergic neurons derived from iPS and ES cells in rat models of Parkinson's disease [35–37] have shown that in vitro-generated dopaminergic neurons can functionally integrate into the brain of adult rats, which suffered from midbrain injury. In all cases, transplanted animals showed improvement in their motor skills. These results underline the clinical potential of iPS cells for personalized cell therapy of Parkinson's disease and other disorders. In order to bring somatic cell reprogramming closer to the clinics, researchers try to find conditions to increase reprogramming efficiencies and to replace virally transduced factors through small compounds and other integration free techniques as summarized in Tables 3.1 and 3.2.

TABLE 3.1 Sources and Methods to Derive iPS cells

Reprogramming Method	Cell Type
Retroviral vectors	Fibroblasts [32] Bone marrow cells [38] Cord blood cells [39] T cells [40] Human umbilical vein endothelial cells [41] Neural stem cells [42] Astrocytes [43] Hepatocytes [44] Amniotic fluid-derived cells [45]
Lentiviral vectors	Peripheral blood cells [46] Cord blood cells [47] T cells [40] Amniotic fluid-derived cells [48]
Cre-recombinase excisable lentiviral vectors	Human dermal fibroblasts [49]
Adenoviral vectors	Mouse primary hepatocytes [50], Human embryonic fibroblasts [51]
Sendai viral vectors	Human neonatal foreskin and adult dermal fibroblasts [52] Human CD34(+) and peripheral blood mononuclear cells [53]
Episomal vectors (derived from the Epstein-Barr virus)	Human foreskin fibroblasts [54]
mRNA	Human neonatal epidermal keratinocytes, human neonatal foreskin fibroblasts, human fetal lung fibroblasts, human fetal skin fibroblasts [55, 56]
Nonviral minicircle vectors	Adipose stem cells [57]
PiggyBac transposon	Mouse and human fibroblasts [58, 59]
miRNA	Mouse adipose stromal cells Human dermal fibroblasts [60]
Proteins	Human neonatal fibroblasts [61]
Liposomal magnetofection	Mouse embryonic fibroblasts [62]

TABLE 3.2 Factors Enhancing iPS Cell Generation

Factors Enhancing Reprogramming	Cell Types
OSKM + vitamin C	Mouse embryonic fibroblasts, human fibroblasts [63]
OSKM + valproic acid, histone deacetylase (HDAC) inhibitors	Mouse embryonic fibroblasts [64]
OSKM, OSM, OSK + sodium butyrate	Human adult and fetal fibroblasts [65]
OSKM + p53 siRNA+ UTF1	Human adult fibroblasts [66]
OSKM + DNMT1 shRNA OSKM + DNMT1 inhibitor 5-aza-cytidine (AZA)	Mouse embryonic fibroblasts [67]
OSKM + Wnt3a	Mouse fibroblasts [68]
OS + Esrrb	Mouse embryonic fibroblasts [69]
OSKM + Pax5 shRNA or C/EBPa	Mouse B cells [70]
OSKM + SB431542, PD0325901, thiazovivin	Human skin fibroblasts [71]
OSKM + Srebp-1	Mouse embryonic fibroblasts [72]
OSKM (or OCT4 alone) + PS48, sodium butyrate, A-83-01, PD0325901	Human neonatal and adult epidermal keratinocytes, human umbilical vein endothelial cells, amniotic fluid-derived cells [73]
OCT4, KLF4 + CHIR99021, Parnate	Mouse embryonic fibroblasts, human neonatal epidermal keratinocytes [74]
OCT4, KLF4 + BIX-01294 and Bay-K-8644	Mouse embryonic fibroblasts [75]

O, Oct4; *S*, Sox2; *K*, Klf4; *M*, c-Myc; *Esrrb*, orphan nuclear receptor; *Srebp-1*, sterol regulatory element-binding protein 1; *PD0325901*, MAPK/ERK pathways inhibitor; *UTF1*, undifferentiated embryonic cell transcription factor 1; *A-83-01*, TGFβ receptor inhibitor; *PS48*, PDK1 (phosphoinositide-dependent protein kinase 1) activator; *Parnate (tranylcypromine)*, inhibitor of lysine-specific demethylase 1; *CHIR99021*, GSK-3 (Glycogen synthase kinase 3) inhibitor; *SB431542*, ALK5 (TGFβ type I receptor kinase) inhibitor; *BIX-01294*, histone methyltransferase (G9aHMTase) inhibitor; *Bay-K-864*, L-type Ca2+ channel activator.

ADULT STEM CELLS

Adult stem cells are multipotent cells that give rise to cell types of one particular tissue to maintain and replenish the function of organs or tissues [76]. Adult stem cells are found in a variety of tissues and organs such as the brain, bone marrow, blood, teeth, heart, gut, and liver [77–83], and are believed to reside in specific niches. In many tissues including the brain and heart, stem cells remain dormant for years until tissue injury, disease, and other factors trigger their activation to replenish the damaged organ. In other tissues such as the bone marrow and the gut, adult stem cells continuously replenish the pool of terminally differentiated cell types.

In contrast to pluripotent stem cells that can differentiate into cell types of all three germ layers, adult stem cells give raise to multiple cell types of one particular tissue. While it is clear that adult stem cells hold great promise for research and therapy, it is also evident that there are limitations that must be overcome in order to consider them ideal candidates for medical applications [84]:

1. It is difficult to maintain and expand functional adult stem cells in vitro, and, in addition, adult stem cells can often be isolated in limited numbers. For therapies, however, adult stem cell should be found in high enough cell numbers to ensure appropriate quantities for further procedures (between 1×10^6 and 1×10^9 cells).
2. Harvest should be doable with minimal invasive procedure(s).
3. Differentiation should be feasible and reliable for multiple cell lineages.
4. Transplantation of these cells has to be safe and effective for autologous and allogeneic transplantations.

Different adult stem cell populations exhibit tissue-specific characteristics and are defined by expression of specific cell surface markers and transcription factors. Here, we will focus on selected adult stem cell types that hold great promise for therapeutic applications.

HEMATOPOIETIC STEM CELLS

Most mature blood cell types are short lived and need to be replenished continuously throughout life. This is guaranteed by the presence of long-term repopulating hematopoietic stem cells (LT-HSC) that reside in the bone marrow niche [85]. In addition to LT-HSC, short-term HSC and progenitor cells can be found in the bone marrow. Marker commonly used to isolate human HSC are CD34+ [86], c-Kit+ [87], Lin-, CD45+ [88], CD90+ and CD49f+ [89], although marker expression varies depending, for instance, on the developmental stage [90].

To release and isolate large numbers of HSC for bone marrow transplantations, cytokine mobilization of HSC into the blood stream is frequently used. Another clinical source for HSC is umbilical cord blood [91–94], which is harvested, frozen, and stored in cord blood banks and can be used as donor-specific or general resource, as needed. HSC isolated from cord blood are used for transplantations in children and, less frequently, in adults. The latter is largely due to the limited number of HSC that can be harvested from each cord.

For both adult and cord blood HSC, there is much interest in identifying the molecular and environmental requirements underlying self-renewal. This would allow ex vivo expansion of functional HSC that can be used for transplantation and that can serve as model system to study human blood development in the dish.

DENTAL STEM CELLS

Dental stem cells (DSCs) are mesenchymal stem cells in the dental pulp. Baby and wisdom teeth as well as healthy molars are considered a great source for DSCs, which have been described as one of the most "powerful stem cell types" in the human body. They replicate faster and for a longer period than adult stem cells derived from other tissues [95]. Similar to cord blood stem cells, dental stem cells can be deposited in stem cell banks. For this, cells are extracted from the dental pulp, processed, and cryopreserved.

The potential of DSCs has been underlined by using them for the reconstruction of human jaw bone using DSCs from wisdom teeth [96, 97]. Another study by Govindasamy et al. [98] isolated DSC from baby teeth to generate islet-like cell aggregates that produce insulin in a glucose-dependent manner. Interestingly, a study from the Nagoya University in Japan demonstrated that DSCs from baby and wisdom teeth can be differentiated into novel neuro-supporting cells that supported healing of spinal cord injuries in rats [95].

ADIPOSE-DERIVED STEM CELLS

Another reliable and easily accessible source of adult stem cell for regenerative medicine is the adipose tissue [99]. Approximately 400,000 liposuction surgeries are performed in the United States each year, yielding up to 3 L of lipoaspirate tissue [100]. This is normally discarded, but could be used for stem cell isolation instead. There is a growing body of evidence from in vitro and in vivo studies that adipose-derived stem cells (ASCs) can be used for a range of therapeutic applications [99] since they are capable of giving rise to adipocytes, chondrocytes, hepatocytes, neuronal cell types, osteoblasts as well as pancreatic and skeletal myocyte lineages (reviewed in detail by Jeffrey M. Gimble [99]).

OLFACTORY ENSHEATING CELLS

Cellular transplantation therapies using different types of cells have been tested in animals and humans to promote functional recovery after spinal cord injury [101]. Among them, transplantation of mucosal olfactory ensheating cells (OECs) represents a promising therapeutic approach for human spinal cord injury owing to high regenerative capacity and autologous origin of OECs [102]. The primary olfactory system, where OECs reside, is unique in supporting neurogenesis throughout life [103]. Moreover, regenerated peripheral axons from the olfactory tissue can enter into the central nervous system environment and form synapses [104]. This unique regenerative capacity of olfactory neurons depends at least partially on specialized features of OECs [105]. In the healthy olfactory tissue, OECs clear axonal debris and dead cells from the system. They also support neurogenesis by reorganizing scar tissue, guiding axons, and secreting neurotrophic factors [106].

Although OECs have a glial nature, they are superior to other glial cells in the central nervous system. In addition to their regenerative potential, it is easier to access OECs as they can be taken from biopsies of olfactory mucosa in the nasal cavity. Further, they can be purified and grown in large numbers. In line with the regenerative potential of OECs, several animal studies showed repair of spinal tract and as a result neurobehavioral/locomotive recovery after OEC transplantation [107–110]. In one of the few human studies mentioned here, OECs isolated from patient's olfactory mucosa biopsy were successfully grown in culture and transplanted into the spinal cord. Transplantation followed by neurorehabilitation resulted in neurological improvements without any adverse effects up to one year after surgery [102]. Another study enrolling small number of trial patients also confirmed the feasibility and safety of OECs purified from nasal biopsies up to three years posttransplantation [111].

One important point that might influence repair efficiency is composition of OEC grafts. For example, it was shown that meningeal cells transplanted together with OECs improve the repair process [112]. Similarly, small molecules, supportive bioactive materials, and additional cells could be implemented together with OECs to increase proliferation and survival of OECs after transplantation, and to enhance neuronal growth and regenerative efficiency.

BIOMATERIALS FOR STEM CELLS

Full application of human pluripotent stem (PS) cells for biomedical applications requires derivation, continuous culturing and differentiation with good manufacturing practices and quality-by-design principles [113–115]. Initially, PS cells were grown and maintained on feeder cells such as mouse embryonic fibroblasts together with culture media and fetal bovine serum. Feeder cells support self-renewal and pluripotency by secreting growth factors, cytokines and extracellular matrix proteins; however, the poorly defined and xenogeneic nature of the feeder cell niche limits the therapeutic use of human PS cells. Feeder-free propagation of human PS cells involves tissue culture plates coated with extracellular matrices such as Matrigel [116]. Because Matrigel is isolated from Engelbroth-Holm-Swarm mouse sarcomas [117] and there is batch-to-batch variability, its use in human biomedical applications is also limited. Alternatives to matrigel are defined and xeno-free substrates such as recombinant human laminin [118] and E-cadherin [119]. In addition to cell-culture substrates, culture media has been developed to support feeder-free culture of human PS cells [120–123]. Chemically defined media can be used together with surfaces coated with xenogeneic-free biological substrates such as vitronectin [123] to improve clinical compliance of human PS cells.

Biological substrates are helpful in biomedical research to study potential applications of human PS cells. However, their clinical use it limited because they show batch-to-batch variability and are a potential source of pathogens as they are not amenable to common sterilization techniques. Synthetic materials such as arrays of laminin-derived peptides with a uniform density and defined orientation [124] or fully synthetic polymer-based substrates [125–128] have been developed to sustain long-term culture of human PS cells. Other approaches include the use of hybrid coatings using polymers together with biomolecules to support self-renewal and differentiation of human ES cells [129]. Taken together, these approaches represent a key step toward long-term maintenance of pluripotent cells in a fully defined microenvironment to accelerate its therapeutic use. Further characterization of chemical and mechanical properties of the synthetic surfaces supporting human PS cell culture is however necessary to develop biomedical applications of these cells.

ORGANOID TECHNOLOGY

Until recently, human ES and iPS cells have been differentiated as monolayer cultures to model human development and disease. Three-dimensional (3D) culture technology show that more complex structures can be grown to better mimic organogenesis in a more physiologically relevant system. These 3D structures, called organoids, were initially developed three decades ago to study organogenesis by cell dissociation and reaggregation experiments. *Re*-emerged organoid technology now describes the platform to generate 3D collection of organ-specific cell types from stem cells in a process recapitulating in vivo self-organization characteristics [130]. Human organoids can be generated from both ES and iPS cells as well as organ-restricted adult stem cells. Extraordinary self-organization potential of stem cells through cell-sorting and spatially restricted lineage determination forms the foundation of organoid development methods. Self-assembly and differentiation of stem cells to form organoids are instructed by extracellular matrix surrounding the cells, the medium, and at later stages by the cells that are present in organoids themselves [131]. Human PS cell-derived organoids have been generated for several organs including gut [132], small intestine [133], kidney [134], pancreas [135], liver [136], brain [137], and retina [138].

Organoid technology is a good model to understand and analyze human embryonic and fetal development. Currently, research on human embryos is not permitted beyond 14 days due to ethical concerns [139]. Therefore, organoid cultures will allow studying human development ex vivo beyond this time point. For example, developing gastric organoids form gland- and pit-like domains, progenitor cells, mucous cells, and endocrine cells and allow to model the development of a stomach [132]. Another example is lung organoids which mimic lung organogenesis with airway-like epithelium, smooth muscle cells, and alveolar-like domain [140]. Moreover, these lung organoids are transcriptionally similar to human fetal lung, suggesting similarity of ex vivo organoid culture and in vivo development [140]. In addition, organoid culture provides a system to study biological paradigms that cannot be modeled in mice due to developmental differences.

Organoid technology also advances biomedical research as they provide model systems to study disorders affecting organs that are difficult to reach, such as the brain. Cell lines do not reflect the morphological and physiological characteristics of neurons, and animal models do not always reflect the true nature of human neurodevelopmental or neurodegenerative diseases. Therefore, organoid culture might be an alternative system to understand human disorders better. In line with this, recent studies elucidated that cerebral organoids can be used to model and study Zika virus exposure [141, 142]. Moreover, transplanted pancreatic organoids with oncogenic mutations were shown to progress to form tumors and led to identification of novel target pathways in pancreatic cancer [135].

Organoid technology offers a platform to test drugs and advance personalized medicine. Since organoid cultures allow modeling human specific tissue structure and disease traits, they can offer better insight into the efficiency and effects of drugs on organs. Improved organoid technology may also affect the process of clinical trials. Preclinical screening of drug safety, efficacy, and therapeutic window can be tested in organoid cultures before proceeding into clinical trials. Furthermore, a selection of patients can be based on their response to compounds tested on organoid cultures generated from the patients' own cells.

Lastly, growing organs for regenerative medicine can be seen as the future of organoid technology. Development of human 3D cultures that resemble specific organs can be used as a source for cell therapies and organ transplantations. Autologous cell transplantation therapies using personalized organoids will circumvent problems related to immunogenicity of transplanted organs. On the other hand, current concerns about potential therapeutic use of organoid technology relate to efficacy and safety as they are derived from pluripotent stem cells. Although organoids generated from adult stem cells have been shown to be genetically stable over time [143], genetic stability, overgrowing, and carcinogenicity of pluripotent stem cell-derived organoids need to be addressed. Furthermore, heterogeneity of generated organoids limits their efficacy in regenerative approaches. Although organoids form tissue layers resembling in vivo counterparts, each organoid formed has a unique organization. One of the reasons for the relatively random positioning of tissue regions in organoid cultures is the lack of embryonic axis formation [130]. This feature makes it for now difficult to generate a reliable source of tissues. Despite these current limitations, however, organoid technology together with CRISPR/Cas9 gene editing technology offers an entirely novel avenue for regenerative medicine.

POTENTIAL APPLICATIONS OF STEM CELLS

As versatile as stem cells are, so are their potential applications. iPS cell biobanks are established to provide study cohorts for research projects, including for example i) the biobank from the National Taiwan University Hospital. Here, blood samples are collected to produce iPS cells from healthy donors and patients suffering from autism spectrum disorders (see NCT02720939) with the aim to generate neural progenitors and neurons and to investigate potential differences, ii) the National Heart Centre Singapore, which is collecting skin biopsies to model the Marfan Syndrome (see NCT02815072), and iii) the Human Induced Pluripotent Stem Cells Initiative (HipSci), a national project in the United Kingdom where iPS cell lines from over 500 individuals with a healthy or a genetic disorder background are generated (see http://www.hipsci.org/).

In addition, over 5700 clinical studies are registered or linked to the term "stem cell" on clinicaltrials.org. These include the investigation of mesenchymal stem cells from adipose tissue for bone fractures in 40 patients (NCT02140528), and the auto- or allogenic transplantation of limbal stem cells for cornea replacement (NCT02318485). The latter is based on a successful phase I/II study, which led to a significant reduced neovascularization in the cornea as well as improved corneal opacity [144].

FUTURE PERSPECTIVES

Improved and refined techniques in isolation and culturing of stem cells will surely improve our knowledge of these cells and bring routine stem cell-based therapies closer to the clinics. One example might be the regulation of stem cell niches, which is important for culture systems and for controlling and limiting self-renewal after transplantation. Moreover, recent advances in gene editing using nucleases (e.g., by CRISPR/Cas9) allow to correct disease-causing mutations. This in turn will lead to novel therapeutic approaches for a variety of malignancies, including the restoration of protein function in disorders such as cystic fibrosis and immunodeficiency. One important aspect for developing therapeutic approaches is the careful evaluation of the long-term behavior of transplanted stem cells. In addition, their potential tumorigenic potential due to self-renewal needs to be monitored. For a broad application of stem cell therapies, efficient protocols allowing to test stem cells prior to transplantation will be crucial.

REFERENCES

[1] Bradley A, Evans M, Kaufman MH, Robertson E. Formation of germ-line chimaeras from embryo-derived teratocarcinoma cell lines. Nature 1984;309(5965):255–6.

[2] Evans MJ, Kaufman MH. Establishment in culture of pluripotential cells from mouse embryos. Nature 1981;292(5819):154–6.

[3] Martin GR. Isolation of a pluripotent cell line from early mouse embryos cultured in medium conditioned by teratocarcinoma stem cells. Proc Natl Acad Sci U S A 1981;78(12):7634–8.

[4] Smith AG, Heath JK, Donaldson DD, Wong GG, Moreau J, Stahl M, et al. Inhibition of pluripotential embryonic stem cell differentiation by purified polypeptides. Nature 1988;336(6200):688–90.

[5] Williams RL, Hilton DJ, Pease S, Willson TA, Stewart CL, Gearing DP, et al. Myeloid leukaemia inhibitory factor maintains the developmental potential of embryonic stem cells. Nature 1988;336(6200):684–7.

[6] Burdon T, Stracey C, Chambers I, Nichols J, Smith A. Suppression of SHP-2 and ERK signalling promotes self-renewal of mouse embryonic stem cells. Dev Biol 1999;210(1):30–43.

[7] Hanna J, Markoulaki S, Mitalipova M, Cheng AW, Cassady JP, Staerk J, et al. Metastable pluripotent states in NOD-mouse-derived ESCs. Cell Stem Cell 2009;4(6):513–24.
[8] Kiyonari H, Kaneko M, Abe S, Aizawa S. Three inhibitors of FGF receptor, ERK, and GSK3 establishes germline-competent embryonic stem cells of C57BL/6N mouse strain with high efficiency and stability. Genesis 2010;48(5):317–27.
[9] Nichols J, Jones K, Phillips JM, Newland SA, Roode M, Mansfield W, et al. Validated germline-competent embryonic stem cell lines from nonobese diabetic mice. Nat Med 2009;15(7):814–8.
[10] Cockburn K, Rossant J. Making the blastocyst: lessons from the mouse. J Clin Invest 2010;120(4):995–1003.
[11] Chazaud C, Yamanaka Y, Pawson T, Rossant J. Early lineage segregation between epiblast and primitive endoderm in mouse blastocysts through the Grb2-MAPK pathway. Dev Cell 2006;10(5):615–24.
[12] Nichols J, Smith A. Naive and primed pluripotent states. Cell Stem Cell 2009;4(6):487–92.
[13] Nichols J, Smith A. The origin and identity of embryonic stem cells. Development 2011;138(1):3–8.
[14] Brons IGM, Smithers LE, Trotter MWB, Rugg-Gunn P, Sun B, Chuva de Sousa Lopes SM, et al. Derivation of pluripotent epiblast stem cells from mammalian embryos. Nature 2007;448(7150):191–5.
[15] Tesar PJ, Chenoweth JG, Brook FA, Davies TJ, Evans EP, Mack DL, et al. New cell lines from mouse epiblast share defining features with human embryonic stem cells. Nature 2007;448(7150):196–9.
[16] Rossant J. Stem cells and early lineage development. Cell 2008;132(4):527–31.
[17] Thomson JA, Itskovitz-Eldor J, Shapiro SS, Waknitz MA, Swiergiel JJ, Marshall VS, et al. Embryonic stem cell lines derived from human blastocysts. Science 1998;282(5391):1145–7.
[18] Shen Y, Matsuno Y, Fouse SD, Rao N, Root S, Xu R, et al. X-inactivation in female human embryonic stem cells is in a nonrandom pattern and prone to epigenetic alterations. Proc Natl Acad Sci 2008;105(12):4709–14.
[19] Sahakyan A, Kim R, Chronis C, Sabri S, Bonora G, Theunissen TW, et al. Human naive pluripotent stem cells model X chromosome dampening and X inactivation. Cell Stem Cell 2017;20(1):87–101.
[20] Patel S, Bonora G, Sahakyan A, Kim R, Chronis C, Langerman J, et al. Human embryonic stem cells do not change their X inactivation status during differentiation. Cell Rep 2017;18(1):54–67.
[21] Nichols J, Smith A. Pluripotency in the embryo and in culture. Cold Spring Harb Perspect Biol 2012;4(8).
[22] Takashima Y, Guo G, Loos R, Nichols J, Ficz G, Krueger F, et al. Resetting transcription factor control circuitry toward ground-state pluripotency in human. Cell 2014;158(6):1254–69.
[23] Theunissen Thorold W, Powell Benjamin E, Wang H, Mitalipova M, Faddah Dina A, Reddy J, et al. Systematic identification of culture conditions for induction and maintenance of naive human pluripotency. Cell Stem Cell 2014;15(4):471–87.
[24] Theunissen Thorold W, Friedli M, He Y, Planet E, O'Neil Ryan C, Markoulaki S, et al. Molecular criteria for defining the naive human pluripotent state. Cell Stem Cell 2016;19(4):502–15.
[25] Ying Q-L, Wray J, Nichols J, Batlle-Morera L, Doble B, Woodgett J, et al. The ground state of embryonic stem cell self-renewal. Nature 2008;453(7194):519–23.
[26] de Wert G, Mummery C. Human embryonic stem cells: research, ethics and policy. Hum Reprod 2003;18(4):672–82.

[27] Lo B, Parham L. Ethical issues in stem cell research. Endocr Rev 2009;30(3):204–13.
[28] Wilmut I, Schnieke AE, McWhir J, Kind AJ, Campbell KH. Viable offspring derived from fetal and adult mammalian cells. Nature 1997;385(6619):810–3.
[29] Hochedlinger K, Jaenisch R. Monoclonal mice generated by nuclear transfer from mature B and T donor cells. Nature 2002;415(6875):1035–8.
[30] Collas P, Taranger CK. Epigenetic reprogramming of nuclei using cell extracts. Stem Cell Rev 2006;2(4):309–17.
[31] Takahashi K, Yamanaka S. Induction of pluripotent stem cells from mouse embryonic and adult fibroblast cultures by defined factors. Cell 2006;126(4):663–76.
[32] Takahashi K, Tanabe K, Ohnuki M, Narita M, Ichisaka T, Tomoda K, et al. Induction of pluripotent stem cells from adult human fibroblasts by defined factors. Cell 2007;131(5):861–72.
[33] Holmqvist S, Lehtonen Š, Chumarina M, Puttonen KA, Azevedo C, Lebedeva O, et al. Creation of a library of induced pluripotent stem cells from Parkinsonian patients. npj Parkinson's Disease 2016;2:16009.
[34] Devine MJ, Ryten M, Vodicka P, Thomson AJ, Burdon T, Houlden H, et al. Parkinson's disease induced pluripotent stem cells with triplication of the alpha-synuclein locus. Nat Commun 2011;2:440.
[35] Rhee YH, Ko JY, Chang MY, Yi SH, Kim D, Kim CH, et al. Protein-based human iPS cells efficiently generate functional dopamine neurons and can treat a rat model of Parkinson disease. J Clin Invest 2011;121(6):2326–35.
[36] Swistowski A, Peng J, Liu QY, Mali P, Rao MS, Cheng LZ, et al. Efficient generation of functional dopaminergic neurons from human induced pluripotent stem cells under defined conditions. Stem Cells 2010;28(10):1893–904.
[37] Grealish S, Diguet E, Kirkeby A, Mattsson B, Heuer A, Bramoulle Y, et al. Human ESC-derived dopamine neurons show similar preclinical efficacy and potency to fetal neurons when grafted in a rat model of Parkinson's disease. Cell Stem Cell 2014;15(5):653–65.
[38] Kunisato A, Wakatsuki M, Kodama Y, Shinba H, Ishida I, Nagao K. Generation of induced pluripotent stem cells by efficient reprogramming of adult bone marrow cells. Stem Cells Dev 2010;19(2):229–38.
[39] Giorgetti A, Montserrat N, Rodriguez-Piza I, Azqueta C, Veiga A, Izpisua Belmonte JC. Generation of induced pluripotent stem cells from human cord blood cells with only two factors: Oct4 and Sox2. Nat Protoc 2010;5(4):811–20.
[40] Loh YH, Hartung O, Li H, Guo C, Sahalie JM, Manos PD, et al. Reprogramming of T cells from human peripheral blood. Cell Stem Cell 2010;7(1):15–9.
[41] Lagarkova MA, Shutova MV, Bogomazova AN, Vassina EM, Glazov EA, Zhang P, et al. Induction of pluripotency in human endothelial cells resets epigenetic profile on genome scale. Cell Cycle 2010;9(5):937–46.
[42] Kim JB, Greber B, Arauzo-Bravo MJ, Meyer J, Park KI, Zaehres H, et al. Direct reprogramming of human neural stem cells by OCT4. Nature 2009;461(7264):649–53.
[43] Ruiz S, Brennand K, Panopoulos AD, Herrerias A, Gage FH, Izpisua-Belmonte JC. High-efficient generation of induced pluripotent stem cells from human astrocytes. PLoS One 2010;5(12):e15526.
[44] Liu H, Ye Z, Kim Y, Sharkis S, Jang YY. Generation of endoderm-derived human induced pluripotent stem cells from primary hepatocytes. Hepatology 2010;51(5):1810–9.
[45] Li C, Zhou J, Shi G, Ma Y, Yang Y, Gu J, et al. Pluripotency can be rapidly and efficiently induced in human amniotic fluid-derived cells. Hum Mol Genet 2009;18(22):4340–9.
[46] Staerk J, Dawlaty MM, Gao Q, Maetzel D, Hanna J, Sommer CA, et al. Reprogramming of human peripheral blood cells to induced pluripotent stem cells. Cell Stem Cell 2010;7(1):20–4.

[47] Haase A, Olmer R, Schwanke K, Wunderlich S, Merkert S, Hess C, et al. Generation of induced pluripotent stem cells from human cord blood. Cell Stem Cell 2009;5(4):434–41.
[48] Zhao HX, Li Y, Jin HF, Xie L, Liu C, Jiang F, et al. Rapid and efficient reprogramming of human amnion-derived cells into pluripotency by three factors OCT4/SOX2/NANOG. Differentiation 2010;80(2–3):123–9.
[49] Soldner F, Hockemeyer D, Beard C, Gao Q, Bell GW, Cook EG, et al. Parkinson's disease patient-derived induced pluripotent stem cells free of viral reprogramming factors. Cell 2009;136(5):964–77.
[50] Okita K, Nakagawa M, Hong HJ, Ichisaka T, Yamanaka S. Generation of mouse induced pluripotent stem cells without viral vectors. Science 2008;322(5903):949–53.
[51] Zhou WB, Freed CR. Adenoviral gene delivery can reprogram human fibroblasts to induced pluripotent stem cells. Stem Cells 2009;27(11):2667–74.
[52] Fusaki N, Ban H, Nishiyama A, Saeki K, Hasegawa M. Efficient induction of transgene-free human pluripotent stem cells using a vector based on Sendai virus, an RNA virus that does not integrate into the host genome. Proc Jpn Acad Ser B Phys Biol Sci 2009;85(8):348–62.
[53] Ye L, Muench MO, Fusaki N, Beyer AI, Wang J, Qi Z, et al. Blood cell-derived induced pluripotent stem cells free of reprogramming factors generated by Sendai viral vectors. Stem Cells Transl Med 2013;2(8):558–66.
[54] Yu JY, Hu KJ, Smuga-Otto K, Tian SL, Stewart R, Slukvin II, et al. Human induced pluripotent stem cells free of vector and transgene sequences. Science 2009;324(5928):797–801.
[55] Warren L, Manos PD, Ahfeldt T, Loh YH, Li H, Lau F, et al. Highly efficient reprogramming to pluripotency and directed differentiation of human cells with synthetic modified mRNA. Cell Stem Cell 2010;7(5):618–30.
[56] Plews JR, Li JL, Jones M, Moore HD, Mason C, Andrews PW, et al. Activation of pluripotency genes in human fibroblast cells by a novel mRNA based approach. Plos One 2010;5(12).
[57] Jia FJ, Wilson KD, Sun N, Gupta DM, Huang M, Li ZJ, et al. A nonviral minicircle vector for deriving human iPS cells. Nat Methods 2010;7(3):197–9.
[58] Kaji K, Norrby K, Paca A, Mileikovsky M, Mohseni P, Woltjen K. Virus-free induction of pluripotency and subsequent excision of reprogramming factors. Nature 2009;458(7239):771–5.
[59] Woltjen K, Michael IP, Mohseni P, Desai R, Mileikovsky M, Hamalainen R, et al. piggyBac transposition reprograms fibroblasts to induced pluripotent stem cells. Nature 2009;458(7239):766–70.
[60] Miyoshi N, Ishii H, Nagano H, Haraguchi N, Dewi DL, Kano Y, et al. Reprogramming of mouse and human cells to pluripotency using mature MicroRNAs. Cell Stem Cell 2011;8(6):633–8.
[61] Kim D, Kim CH, Moon JI, Chung YG, Chang MY, Han BS, et al. Generation of human induced pluripotent stem cells by direct delivery of reprogramming proteins. Cell Stem Cell 2009;4(6):472–6.
[62] Park HY, Noh EH, Chung HM, Kang MJ, Kim EY, Park SP. Efficient generation of virus-free iPS cells using liposomal magnetofection. PLoS One 2012;7(9).
[63] Esteban MA, Wang T, Qin B, Yang J, Qin D, Cai J, et al. Vitamin C enhances the generation of mouse and human induced pluripotent stem cells. Cell Stem Cell 2010;6(1):71–9.
[64] Huangfu DW, Maehr R, Guo WJ, Eijkelenboom A, Snitow M, Chen AE, et al. Induction of pluripotent stem cells by defined factors is greatly improved by small-molecule compounds. Nat Biotechnol 2008;26(7):795–7.
[65] Mali P, Chou BK, Yen J, Ye Z, Zou J, Dowey S, et al. Butyrate greatly enhances derivation of human induced pluripotent stem cells by promoting epigenetic remodeling and the expression of pluripotency-associated genes. Stem Cells 2010;28(4):713–20.

[66] Zhao Y, Yin XL, Qin H, Zhu FF, Liu HS, Yang WF, et al. Two supporting factors greatly improve the efficiency of human iPSC generation. Cell Stem Cell 2008;3(5):475–9.
[67] Mikkelsen TS, Hanna J, Zhang XL, Ku MC, Wernig M, Schorderet P, et al. Dissecting direct reprogramming through integrative genomic analysis. Nature 2008;454(7200):49–55.
[68] Marson A, Foreman R, Chevalier B, Bilodeau S, Kahn M, Young RA, et al. Wnt signaling promotes reprogramming of somatic cells to pluripotency. Cell Stem Cell 2008;3(2):132–5.
[69] Feng B, Jiang JM, Kraus P, Ng JH, Heng JCD, Chan YS, et al. Reprogramming of fibroblasts into induced pluripotent stem cells with orphan nuclear receptor Esrrb. Nat Cell Biol 2009;11(2):197–203.
[70] Hanna J, Markoulaki S, Schorderet P, Carey BW, Beard C, Wernig M, et al. Direct reprogramming of terminally differentiated mature B lymphocytes to pluripotency (vol 133, pg 250, 2008). Cell 2008;134(2):365.
[71] Lin T, Ambasudhan R, Yuan X, Li W, Hilcove S, Abujarour R, et al. A chemical platform for improved induction of human iPSCs. Nat Methods 2009;6(11):805–8.
[72] Wu Y, Chen KS, Liu XY, Huang LL, Zhao DY, Li LP, et al. Srebp-1 interacts with c-Myc to enhance somatic cell reprogramming. Stem Cells 2016;34(1):83–92.
[73] Zhu SY, Li WL, Zhou HY, Wei WG, Ambasudhan R, Lin TX, et al. Reprogramming of human primary somatic cells by OCT4 and chemical compounds. Cell Stem Cell 2010;7(6):651–5.
[74] Li WL, Zhou HY, Abujarour R, Zhu SY, Joo JY, Lin TX, et al. Generation of human-induced pluripotent stem cells in the absence of exogenous Sox2. Stem Cells 2009;27(12):2992–3000.
[75] Shi Y, Desponts C, Do JT, Hahm HS, Scholer HR, Ding S. Induction of pluripotent stem cells from mouse embryonic fibroblasts by Oct4 and Klf4 with small-molecule compounds. Cell Stem Cell 2008;3(5):568–74.
[76] Wagers AJ, Weissman IL. Plasticity of adult stem cells. Cell 2004;116(5):639–48.
[77] Reynolds BA, Weiss S. Generation of neurons and astrocytes from isolated cells of the adult mammalian central nervous system. Science 1992;255(5052):1707–10.
[78] Wognum AW, Eaves AC, Thomas TE. Identification and isolation of hematopoietic stem cells. Arch Med Res 2003;34(6):461–75.
[79] Coles BL, van der Kooy D. Isolation of retinal stem cells from the mouse eye. J Vis Exp 2010;(43).
[80] Vishwanath VR, Nadig RR, Nadig R, Prasanna JS, Karthik J, Pai VS. Differentiation of isolated and characterized human dental pulp stem cells and stem cells from human exfoliated deciduous teeth: an in vitro study. J Conserv Dent 2013;16(5):423–8.
[81] Smith AJ, Lewis FC, Aquila I, Waring CD, Nocera A, Agosti V, et al. Isolation and characterization of resident endogenous c-kit+ cardiac stem cells from the adult mouse and rat heart. Nat Protoc 2014;9(7):1662–81.
[82] Haegebarth A, Clevers H. Wnt signaling, lgr5, and stem cells in the intestine and skin. Am J Pathol 2009;174(3):715–21.
[83] Herrera MB, Bruno S, Buttiglieri S, Tetta C, Gatti S, Deregibus MC, et al. Isolation and characterization of a stem cell population from adult human liver. Stem Cells 2006;24(12):2840–50.
[84] Gimble JM. Adipose tissue-derived therapeutics. Expert Opin Biol Ther 2003;3(5):705–13.
[85] Ackermann M, Liebhaber S, Klusmann JH, Lachmann N. Lost in translation: pluripotent stem cell-derived hematopoiesis. EMBO Mol Med 2015;7(11):1388–402.
[86] Sidney LE, Branch MJ, Dunphy SE, Dua HS, Hopkinson A. Concise review: evidence for CD34 as a common marker for diverse progenitors. Stem Cells 2014;32(6):1380–9.
[87] Ikuta K, Weissman IL. Evidence that hematopoietic stem cells express mouse c-kit but do not depend on steel factor for their generation. Proc Natl Acad Sci U S A 1992;89(4):1502–6.

[88] Craig W, Poppema S, Little MT, Dragowska W, Lansdorp PM. CD45 isoform expression on human haemopoietic cells at different stages of development. Br J Haematol 1994;88(1): 24–30.

[89] Notta F, Doulatov S, Laurenti E, Poeppl A, Jurisica I, Dick JE. Isolation of single human hematopoietic stem cells capable of long-term multilineage engraftment. Science 2011;333(6039):218–21.

[90] McKinney-Freeman SL, Naveiras O, Yates F, Loewer S, Philitas M, Curran M, et al. Surface antigen phenotypes of hematopoietic stem cells from embryos and murine embryonic stem cells. Blood 2009;114(2):268–78.

[91] Broxmeyer HE, Douglas GW, Hangoc G, Cooper S, Bard J, English D, et al. Human umbilical-cord blood as a potential source of transplantable hematopoietic stem progenitor cells. Proc Natl Acad Sci U S A 1989;86(10):3828–32.

[92] Takahashi S, Ooi J, Tomonari A, Konuma T, Fukuno K, Tsukada N, et al. Clinical outcomes of cord blood transplantation from unrelated donors comparable with marrow or blood transplantation from related donors in adults: a single institute analysis. Blood 2005;106(11):93.

[93] Gluckman E, Rocha V, BoyerChammard A, Locatelli F, Arcese W, Pasquini R, et al. Outcome of cord-blood transplantation from related and unrelated donors. N Engl J Med 1997;337(6):373–81.

[94] Smith AR, Wagner JE. Alternative haematopoietic stem cell sources for transplantation: place of umbilical cord blood. Br J Haematol 2009;147(2):246–61.

[95] Sakai K, Yamamoto A, Matsubara K, Nakamura S, Naruse M, Yamagata M, et al. Human dental pulp-derived stem cells promote locomotor recovery after complete transection of the rat spinal cord by multiple neuro-regenerative mechanisms. J Clin Investig 2012;122(1): 80–90.

[96] Giuliani A, Manescu A, Langer M, Rustichelli F, Desiderio V, Paino F, et al. Three years after transplants in human mandibles, histological and in-line holotomography revealed that stem cells regenerated a compact rather than a spongy bone: biological and clinical implications. Stem Cells Transl Med 2013;2(4):316–24.

[97] d'Aquino R, De Rosa A, Lanza V, Tirino V, Laino L, Graziano A, et al. Human mandible bone defect repair by the grafting of dental pulp stem/progenitor cells and collagen sponge biocomplexes. Eur Cell Mater 2009;18:75–83.

[98] Govindasamy V, Ronald VS, Abdullah AN, Nathan KRG, Aziz ZACA, Abdullah M, et al. Differentiation of dental pulp stem cells into islet-like aggregates. J Dent Res 2011;90(5): 646–52.

[99] Gimble JM, Katz AJ, Bunnell BA. Adipose-derived stem cells for regenerative medicine. Circ Res 2007;100(9):1249–60.

[100] Katz AJ, Llull R, Hedrick MH, Futrell JW. Emerging approaches to the tissue engineering of fat. Clin Plast Surg 1999;26(4):587–603. [viii].

[101] Tetzlaff W, Okon EB, Karimi-Abdolrezaee S, Hill CE, Sparling JS, Plemel JR, et al. A systematic review of cellular transplantation therapies for spinal cord injury. J Neurotrauma 2011;28(8):1611–82.

[102] Tabakow P, Jarmundowicz W, Czapiga B, Fortuna W, Miedzybrodzki R, Czyz M, et al. Transplantation of autologous olfactory ensheathing cells in complete human spinal cord injury. Cell Transplant 2013;22(9):1591–612.

[103] Farbman AI. Olfactory neurogenesis: genetic or environmental controls? Trends Neurosci 1990;13(9):362–5.

[104] Doucette R. Glial influences on axonal growth in the primary olfactory system. Glia 1990;3(6):433–49.

[105] Barnett SC, Riddell JS. Olfactory ensheathing cells (OECs) and the treatment of CNS injury: advantages and possible caveats. J Anat 2004;204(1):57–67.
[106] Roet KCD, Verhaagen J. Understanding the neural repair-promoting properties of olfactory ensheathing cells. Exp Neurol 2014;261:594–609.
[107] López-Vales R, Forés J, Navarro X, Verdú E. Chronic transplantation of olfactory ensheathing cells promotes partial recovery after complete spinal cord transection in the rat. Glia 2007;55(3):303–11.
[108] Watzlawick R, Rind J, Sena ES, Brommer B, Zhang T, Kopp MA, et al. Olfactory Ensheathing cell transplantation in experimental spinal cord injury: effect size and reporting Bias of 62 experimental treatments: a systematic review and meta-analysis. PLoS Biol 2016;14(5):e1002468.
[109] Lu J, Féron F, Mackay-Sim A, Waite PME. Olfactory ensheathing cells promote locomotor recovery after delayed transplantation into transected spinal cord. Brain 2002;125(1):14–21.
[110] Li Y, Field PM, Raisman G. Repair of adult rat corticospinal tract by transplants of olfactory ensheathing cells. Science 1997;277(5334):2000–2.
[111] Mackay-Sim A, Féron F, Cochrane J, Bassingthwaighte L, Bayliss C, Davies W, et al. Autologous olfactory ensheathing cell transplantation in human paraplegia: a 3-year clinical trial. Brain 2008;131(9):2376–86.
[112] Lakatos A, Smith PM, Barnett SC, Franklin RJM. Meningeal cells enhance limited CNS remyelination by transplanted olfactory ensheathing cells. Brain 2003;126(3):598–609.
[113] Villa-Diaz LG, Ross AM, Lahann J, Krebsbach PH. Concise review: the evolution of human pluripotent stem cell culture: from feeder cells to synthetic coatings. Stem Cells 2013;31(1):1–7.
[114] Celiz AD, Smith JGW, Langer R, Anderson DG, Winkler DA, Barrett DA, et al. Materials for stem cell factories of the future. Nat Mater 2014;13(6):570–9.
[115] Lipsitz YY, Timmins NE, Zandstra PW. Quality cell therapy manufacturing by design. Nat Biotechnol 2016;34(4):393–400.
[116] Xu C, Inokuma MS, Denham J, Golds K, Kundu P, Gold JD, et al. Feeder-free growth of undifferentiated human embryonic stem cells. Nat Biotechnol 2001;19(10):971–4.
[117] Hassell JR, Robey PG, Barrach HJ, Wilczek J, Rennard SI, Martin GR. Isolation of a heparan sulfate-containing proteoglycan from basement membrane. Proc Natl Acad Sci U S A 1980;77(8):4494–8.
[118] Miyazaki T, Futaki S, Hasegawa K, Kawasaki M, Sanzen N, Hayashi M, et al. Recombinant human laminin isoforms can support the undifferentiated growth of human embryonic stem cells. Biochem Biophys Res Commun 2008;375(1):27–32.
[119] Nagaoka M, Si-Tayeb K, Akaike T, Duncan SA. Culture of human pluripotent stem cells using completely defined conditions on a recombinant E-cadherin substratum. BMC Dev Biol 2010;10(1):60.
[120] Ludwig TE, Levenstein ME, Jones JM, Berggren WT, Mitchen ER, Frane JL, et al. Derivation of human embryonic stem cells in defined conditions. Nat Biotechnol 2006;24(2): 185–7.
[121] Lu J, Hou R, Booth CJ, Yang S-H, Snyder M. Defined culture conditions of human embryonic stem cells. Proc Natl Acad Sci 2006;103(15):5688–93.
[122] Yao S, Chen S, Clark J, Hao E, Beattie GM, Hayek A, et al. Long-term self-renewal and directed differentiation of human embryonic stem cells in chemically defined conditions. Proc Natl Acad Sci 2006;103(18):6907–12.
[123] Chen G, Gulbranson DR, Hou Z, Bolin JM, Ruotti V, Probasco MD, et al. Chemically defined conditions for human iPSC derivation and culture. Nat Methods 2011;8(5):424–9.

[124] Derda R, Li L, Orner BP, Lewis RL, Thomson JA, Kiessling LL. Defined substrates for human embryonic stem cell growth identified from surface arrays. ACS Chem Biol 2007;2(5):347–55.

[125] Villa-Diaz LG, Nandivada H, Ding J, Nogueira-de-Souza NC, Krebsbach PH, O'Shea KS, et al. Synthetic polymer coatings for long-term growth of human embryonic stem cells. Nat Biotechnol 2010;28(6):581–3.

[126] Mei Y, Saha K, Bogatyrev SR, Yang J, Hook AL, Kalcioglu ZI, et al. Combinatorial development of biomaterials for clonal growth of human pluripotent stem cells. Nat Mater 2010;9(9):768–78.

[127] Nandivada H, Villa-Diaz LG, O'Shea KS, Smith GD, Krebsbach PH, Lahann J. Fabrication of synthetic polymer coatings and their use in feeder-free culture of human embryonic stem cells. Nat Protocols 2011;6(7):1037–43.

[128] Irwin EF, Gupta R, Dashti DC, Healy KE. Engineered polymer-media interfaces for the long-term self-renewal of human embryonic stem cells. Biomaterials 2011;32(29):6912–9.

[129] Melkoumian Z, Weber JL, Weber DM, Fadeev AG, Zhou Y, Dolley-Sonneville P, et al. Synthetic peptide-acrylate surfaces for long-term self-renewal and cardiomyocyte differentiation of human embryonic stem cells. Nat Biotechnol 2010;28(6):606–10.

[130] Lancaster MA, Knoblich JA. Organogenesis in a dish: modeling development and disease using organoid technologies. Science 2014;345(6194).

[131] Huch M, Knoblich JA, Lutolf MP, Martinez-Arias A. The hope and the hype of organoid research. Development 2017;144(6):938–41.

[132] McCracken KW, Cata EM, Crawford CM, Sinagoga KL, Schumacher M, Rockich BE, et al. Modelling human development and disease in pluripotent stem-cell-derived gastric organoids. Nature 2014;516(7531):400–4.

[133] Spence JR, Mayhew CN, Rankin SA, Kuhar MF, Vallance JE, Tolle K, et al. Directed differentiation of human pluripotent stem cells into intestinal tissue in vitro. Nature 2011;470(7332):105–9.

[134] Takasato M, Er PX, Chiu HS, Maier B, Baillie GJ, Ferguson C, et al. Kidney organoids from human iPS cells contain multiple lineages and model human nephrogenesis. Nature 2015;526(7574):564–8.

[135] Boj Sylvia F, Hwang C-I, Baker Lindsey A, Chio Iok In C, Engle Dannielle D, Corbo V, et al. Organoid models of human and mouse ductal pancreatic cancer. Cell 2015;160(1):324–38.

[136] Takebe T, Sekine K, Enomura M, Koike H, Kimura M, Ogaeri T, et al. Vascularized and functional human liver from an iPSC-derived organ bud transplant. Nature 2013;499(7459):481–4.

[137] Lancaster MA, Renner M, Martin C-A, Wenzel D, Bicknell LS, Hurles ME, et al. Cerebral organoids model human brain development and microcephaly. Nature 2013;501(7467):373–9.

[138] Nakano T, Ando S, Takata N, Kawada M, Muguruma K, Sekiguchi K, et al. Self-formation of optic cups and storable stratified neural retina from human ESCs. Cell Stem Cell 2012;10(6):771–85.

[139] Bredenoord AL, Clevers H, Knoblich JA. Human tissues in a dish: the research and ethical implications of organoid technology. Science 2017;355(6322).

[140] Dye BR, Hill DR, Ferguson MAH, Tsai Y-H, Nagy MS, Dyal R, et al. In vitro generation of human pluripotent stem cell derived lung organoids. Elife 2015;4:e05098.

[141] Qian X, Nguyen Ha N, Song Mingxi M, Hadiono C, Ogden Sarah C, Hammack C, et al. Brain-region-specific organoids using mini-bioreactors for modeling ZIKV exposure. Cell 2016;165(5):1238–54.

[142] Dang J, Tiwari SK, Lichinchi G, Qin Y, Patil Veena S, Eroshkin Alexey M, et al. Zika virus depletes neural progenitors in human cerebral organoids through activation of the innate immune receptor TLR3. Cell Stem Cell 2016;19(2):258–65.
[143] Huch M, Gehart H, van Boxtel R, Hamer K, Blokzijl F, Verstegen Monique MA, et al. Long-term culture of genome-stable bipotent stem cells from adult human liver. Cell 2015;160(1–2):299–312.
[144] Zakaria N, Possemiers T, Dhubhghaill SN, Leysen I, Rozema J, Koppen C, et al. Results of a phase I/II clinical trial: standardized, non-xenogenic, cultivated limbal stem cell transplantation. J Transl Med 2014;12.

Chapter 4

Transdifferentiation—Changing Cell Identity

Theresa D. Ahrens[*], Safak Caglayan[*], Judith Staerk[*,†,‡], Artur Cieślar-Pobuda[*]
[*]Nordic European Molecular Laboratory (EMBL) Partnership, Centre for Molecular Medicine Norway, University of Oslo, Oslo, Norway
[†]Department of Haematology, Oslo University Hospital, Oslo, Norway
[‡]Norwegian Center for Stem Cell Research, Department of Immunology, Oslo University Hospital, Oslo, Norway

ABBREVIATIONS

AZA	5-azacytidine
BMPs	bone morphogenetic proteins
cAMP	3′,5′-cyclic adenosine monophosphate
H3K27me3	histone H3 trimethylation on lysine 27
H3K4me3	histone H3 trimethylation on lysine 4
H3K9me	lysine 9 trimethylation at histone H3
iPSC	induced pluripotent stem cell
JNK	c-Jun N-terminal kinases
MI	myocardial infarction
miRNAs	micro-RNAs
Mps1	monopolar spindle 1
MyoD	myogenic differentiation 1
NPCs	neural progenitor cells
Pax6	paired box protein pax-6
PcG	polycomb group
PDX1	pancreas/duodenal homeobox gene 1
Plk1	polo-like kinase 1
PRCs	polycomb-repressive complexes
RITA	reactivation of p53 and induction of tumor apoptosis
TF	transcription factor
TrxG	trithorax group
vmhc	ventricular myosin heavy chain

Stem Cells and Biomaterials for Regenerative Medicine. https://doi.org/10.1016/B978-0-12-812258-7.00004-6

INTRODUCTION

The primary goal of regenerative medicine is to replace damaged tissue that cannot be repaired by the body itself. This is necessary since growing demands for transplantation therapies far outweigh available donor material. Today, regeneration therapy is mainly associated with stem cells, which after differentiation into a tissue or cell type of choice are transplanted back into the patient [1]. It is possible to circumvent the use of stem cells through a process called transdifferentiation. Transdifferentiation, also called direct reprogramming, describes the conversion of one cell type into another without going through the pluripotent cell state and is closely related to "reprogramming" and "dedifferentiation" (Fig. 4.1).

Transdifferentiation is thought to be a valuable alternative for induced pluripotent stem cell (iPSC) technology in cell replacement therapies. By directly converting differentiated cells into other somatic cell types, the state of pluripotency is avoided and the risk of tumorigenesis is thought to be significantly reduced. Another advantage of transdifferentiation is the shorter time required to obtain the cell type of interest. Differentiation of stem cells to the tissue of

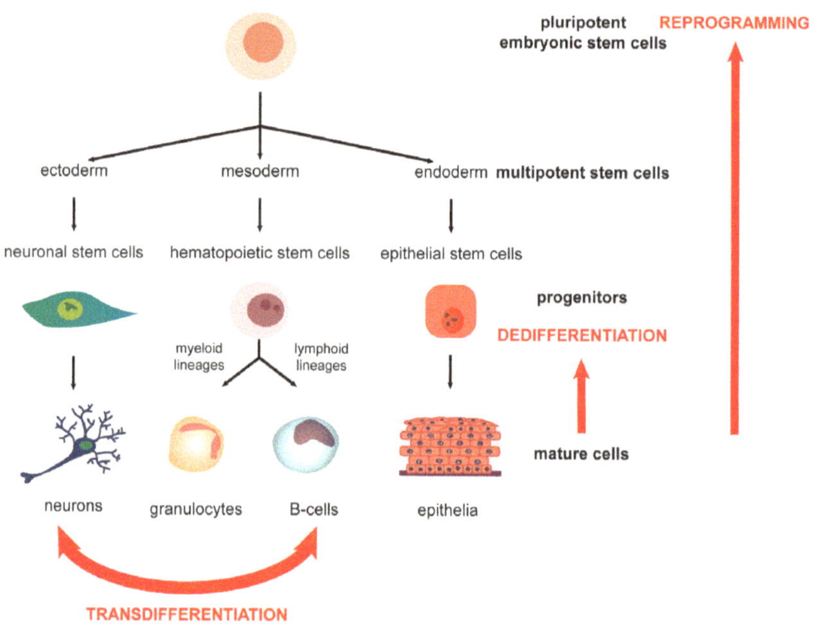

FIG. 4.1 Dedifferentiation, transdifferentiation, and reprogramming. During development, embryonic stem cells (pluripotent) give rise to all mature cell types found in the body. Differentiation can be reversed by (i) dedifferentiation, a naturally occurring process in which somatic cell types reverse to an undifferentiated progenitor cell, (ii) transdifferentiation, a direct change in cell identity without going through the pluripotent stem cell state, and (iii) reprogramming, a technique where pluripotency is induced in somatic cell types.

choice generally takes between two to three weeks, while transdifferentiation takes only a few days. Reprogramming describes the induced conversion of adult somatic cells into pluripotent stem cells, while dedifferentiation is referred to as a process in which somatic cell types reverse to an undifferentiated progenitor cell. This in turn allows the cell to proliferate and differentiate again which will eventually lead to the replacement of cells or tissues that have been lost.

IN VIVO TRANSDIFFERENTIATION: EXAMPLES OF NATURALLY OCCURRING REGENERATION

Dedifferentiation and transdifferentiation are processes that occur naturally during organ regeneration. Regeneration of body parts has been documented as early as the 18th century when Charles Bonnet observed that salamander had the capability to regenerate their eye after it had been removed. Around the same time, it was noted that crayfish can regenerate a new limb after amputation [2]. In 1891 and 1895, regeneration of the eye lens in adult newts was independently described by Colucci [3] and Wolff [4], respectively. Ever since, regeneration processes in animals have been extensively studied. One example of in vivo dedifferentiation can be found during regeneration of limbs in urodele amphibians where amputation of the limb stimulates the cells adjacent to the wound and leads to formation of blastema with undifferentiated cells. These cells then proliferate and differentiate to reconstruct the lost limb [5]. Another example has been described by Sandoval-Guzman who demonstrated that certain species of salamanders, after limb amputation, use dedifferentiation to have their muscle fibers produce muscle progenitors necessary for limb regeneration [6]. In addition, removal of the lens from African clawed frog (*Xenopus laevis*) triggers a transdifferentiation process in which the central corneal cells form a new lens. It is thought that this process is induced by exposure of the cornea to the vitreous humor which takes place after lens removal. The authors demonstrated that such transdifferentiation from cornea to lens depends on activation of *bone morphogenetic proteins* (BMPs) and Wnt [7] signaling pathways.

More recently, planaria, a flatworm, has emerged as model system to study cell plasticity and regeneration processes. Despite a complex anatomy, including for example a central nervous system, amputation of body parts leads to full regeneration independent of the cut axis (longitudinal or transverse) [8–10]. This is due to so-called neoblasts which have been identified to give rise to different tissues and/or cell types during regeneration. Strikingly, transplantation of a single neoblast cell rescues lethally irradiated planaria [11], demonstrating their pluripotent potential. Scimone and colleagues then showed that neoblasts show a lineage determination toward distinct tissues 48 h after amputation [12]. Regeneration in planaria involves two proliferation waves: (i) neoblasts proliferating in the whole body and (ii) neoblasts undergoing locally regulated proliferation and subsequent regeneration [13]. Regeneration by neoblasts is controlled by many different mechanisms such as the WNT/β-catenin pathway that controls the anterior-posterior body axis [14] and is activated by amputation or wounding [15]. Moreover, control of regeneration in planaria involves bioelectrical signals. The H/K-ATPase balances apoptosis and tissue regeneration in such way that regenerated organs have proper size [16]. Certainly, further mechanisms controlling planaria regeneration will be unraveled in future.

Another example of regeneration occurs during heart regeneration in zebrafish [17]. When part of the zebrafish heart ventricle is removed, cardiomyocytes dedifferentiate and increase their proliferation rate. This process is known to be accompanied by upregulation of positive cell cycle regulators, such as polo-like kinase 1 (*plk1*), and monopolar spindle 1 (*mps1*), downregulation of ventricular myosin heavy chain (*vmhc*), and disassembly of the sarcomeric contractile apparatus [18–20]. Zebrafish are also able to regenerate their bones via dedifferentiation of osteoblasts in fins as described by Knopf et al. [21].

Aside from regeneration processes, transdifferentiation can also be part of development. As an important model system to study such processes served the nematode *Caenorhabditis elegans* which is unique with respect to the fact that adult hermaphrodites consist of exactly 959 somatic cells. Detailed cell fate mapping lead to a family tree unraveling the origin and destiny of every single somatic cell. For instance, *C. elegans* shows a unique epithelial-neuronal transdifferentiation during development where the epithelial cell called "Y" is part of the rectum, starts migrating and transforms into a motoneuron called "PDA." This process is associated with decreased expression of epithelial markers and progressive increase of neuronal markers [22]. Studies on this embryonic transdifferentiation process reveal that this conversion is a multistep process with defined gene expression states and without going through a multipotent stem cell-like state [23].

Naturally occurring regeneration has also been observed in mammals, including humans. Well-documented examples are myelinating Schwann cells [24, 25] that after being damaged are able to dedifferentiate, re-enter the cell cycle, and start proliferating. Although the underlying molecular events of this process are not well understood, it has been shown that 3′,5′-cyclic adenosine monophosphate (cAMP) is responsible for initiation and maintaining a state of differentiation. Addition and removal of cAMP to isolated Schwann cells induces multiple cycles of differentiation and dedifferentiation. At the same time, it has been shown that activation of receptor tyrosine kinases by different growth factors is not sufficient for triggering dedifferentiation in the presence of cAMP, although removal of cAMP initiates the dedifferentiation process through a mechanism that requires c-Jun N-terminal kinase (JNK) [25].

Cellular transdifferentiation is also known to naturally occur during liver regeneration and in the pancreas. Yanger et al. [26] showed that in order to induce transdifferentiation of hepatocytes into biliary cells, injury or forced NOTCH signaling is sufficient. Furthermore, it has been demonstrated that after extreme loss of pancreatic β-cells, α-cells can transdifferentiate into functional β-cells and trigger their own transdifferentiation without the need of external factors [27]. Cells resembling functional β-cells can also be obtained by forcing expression of transcription factors (TFs) (see "Therapeutic Transdifferentiation" section for more details).

Studying naturally occurring regenerative processes are crucial to define molecular and cellular mechanisms underlying transdifferentiation, translate them into in vitro studies and eventually use them to develop regenerative strategies for humans [28]. In recent years, much progress has been made to define how one cell type can be converted to another as outlined below. Many factors required to change cell identity have been identified and successful cellular conversions have been performed. Nevertheless, the mechanisms regulating cellular plasticity remain incompletely understood. It is known that changes in gene expression on both, the transcriptional and posttranslational level, are required for transdifferentiation. In addition, these processes need to be precisely regulated in a controlled spatial and temporal manner. The current understanding is that transdifferentiation and dedifferentiation are triggered by expression changes of key TFs that alter the developmental commitment of the cell [1]. It has further been suggested that transdifferentiation may result from somatic mutation of homeotic genes normally required to distinguish tissue rudiments from one another [29]. In any case, chromatin remodeling, the process by which new transcriptional programs are activated or inactivated, is likely to be at the center of this process [30].

It is also well established that cellular mechanisms exist through which tissues can regenerate in response to stem cell ablation. Recently, Tata et al. showed that fully committed secretory cells respond to stem cell ablation by enhancing their proliferation and converting into functional epithelial stem cells. The ability of differentiated cells to regain stem cell properties was found to be inversely proportional to the degree of the maturity of the differentiated cells. Moreover, this dedifferentiation process was shown to be inhibited and regulated by contact with existing stem cells [31].

GENE REGULATION DURING TRANSDIFFERENTIATION

During development, a single totipotent cell (the fertilized oocyte) gives rise to all mature cell types. Cell lineage determination is tightly controlled by signaling pathways and gene regulation through epigenetic changes including DNA methylation, histone marks, and miRNAs. These epigenetic traits work together to establish stable gene expression patterns and are associated with a general regulation of chromatin accessibility. Chromatin can be divided into euchromatin, associated with an open DNA conformation and active gene transcription, and heterochromatin, associated with a closed DNA conformation and gene silencing. Both chromatin states are characterized by specific DNA and/ or histone modifications. Euchromatin has been linked to histone acetylation or lysine 4 trimethylation at histone H3 (H3K4me3) and heterochromatin is associated with lysine 9 trimethylation at histone H3 (H3K9me3) as well as with DNA methylation. This complex network of histone marks is controlled by many different writers, readers, erasers, which orchestrate the histone code to regulate gene expression [32].

Indeed, pluripotent stem cells are characterized by a special chromatin state which is globally decondensed, enriched in active histone marks and transcriptionally hyperactive [33, 34]. During embryonic development and cell fate decisions, activation and inactivation of gene expression is important: gene activation is controlled by the trithorax group proteins (TrxG) and gene repression by the polycomb group (PcG) proteins. The common mechanism of both protein complexes is the control of gene expression via modification of histone marks.

The TrxG family regulates gene expression by setting activating histone marks, such as histone H3K4me3. In contrast, the PcG family forms multi-protein complexes (PRCs) and regulates gene repression through inactivating histone marks, such as histone H3 trimethylation on lysine 27 (H3K27me3) [35, 36]. Indeed, formation of heterochromatin in stem cells is essential to exit the pluripotent state. This heterochromatinization is initiated at the nucleolus and accompanied by increasing levels of repressive histone marks [37]. Moreover, different epigenetic mechanisms cooperate to regulate gene expression. This has been nicely demonstrated for the pluripotency gene *Oct3/4* which becomes epigenetically silenced during differentiation. Here, binding of the histone methyltransferase G9a leads to increased H3K9me3 levels and binding of the heterochromatin protein 1. This in turn leads to the recruitment of the de novo methyltransferases Dnmt3a and Dnmt3b and subsequent promotor DNA hypermethylation and gene silencing [38, 39].

In the beginning it was thought that the differentiation process is irreversible, as has for example been illustrated by Waddington's epigenetic landscape. Meanwhile, however, it is clear that "differentiation" can be a quite fluctuant state. To date three main mechanisms that can reverse differentiation have been defined [5]: dedifferentiation, transdifferentiation, and reprogramming (see Fig. 4.1). All these processes involve reshaping of the chromatin structure and alterations of gene expression patterns.

Active gene transcription is dependent on a network of so-called general TFs which enable gene transcription via RNA polymerase II [40, 41]. For this, DNA binding sites of the TFs have to be accessible in an open chromatin state. In contrast, a special set of so-called pioneer TFs is able to recognize their binding sites in closed heterochromatin [42, 43]. Known pioneer factors are for example FoxA and GATA family TFs as well as PU.1. The exact mechanism underlying chromatin restructuring is incompletely understood. It is thought that pioneering factors are either recruiting other chromatin-modifying complexes to alter and unwrap the (local) chromatin and thus allow gene expression, or that they have intrinsic chromatin remodeling activity. For instance, the pioneer factor FoxA1 is able to release the linker histone H1 from the DNA [44] which is important for higher order and closed chromatin structure.

With this unusual concept of chromatin binding, pioneer TFs are crucial for reprogramming [45] and transdifferentiation [46]. The conversion of fibroblasts into hepatocyte-like cells requires expression of Hnf4a and one pioneer factor (either FoxA1, FoxA2, or FoxA3) [47], while reprogramming into iPSCs has been shown to rely on the core set of four TFs Oct3/4, Sox2, c-Myc, and Klf4 [48]. Three of these reprogramming factors, Oct4, Sox2, and Klf4, have been shown to act as pioneer TFs during reprogramming [49]. Interestingly, these reprogramming factors do not bind to gene promotors where gene transcription is initiated, but to distal DNA regulatory elements called "enhancers" [49, 50] which are known to establish tissue-specific gene expression. This is possible due to long-range interactions and looping of the DNA to bring enhancers and promotors (with the transcription start site) in proximity to each other.

Lee et al. compared a nonintegrating approach of cell-permeable proteins against retroviral overexpression during reprogramming to iPSCs, and showed that retroviral overexpression is more efficient [51]. Indeed, inflammatory and innate immune response genes are only increased by the retroviral overexpression approach, and additional experiments revealed that activation of Toll-like receptor 3 (TLR3) signaling is responsible for the higher efficiency. TLR3 signaling has been demonstrated to reshape the epigenetic landscape on many different levels: histone deacetylases are downregulated which blocks the removal of histone acetylation and leads to a subsequent accumulation of histone acetylation. Histone hyperacetylation is generally associated with an open chromatin conformation which is known to facilitate reprogramming. Based on these results, Lee et al. establish the term "transflammation" to point out that inflammatory processes might support reprogramming processes by influencing/supporting chromatin accessibility [51]. While this hypothesis still needs to be proven for transdifferentiation, it is clear that inflammatory processes are involved in cell differentiation in humans. Chronic inflammation is for example associated with gastroesophageal reflux disease and is linked to the development of Barrett's esophagus, although the exact mechanisms of this conversion are still being discussed [52, 53].

The exact mechanisms underlying transdifferentiation might be different for every cell type and conversion protocol, depending on the distinct and indispensable TFs needed for each unique conversion process. Understanding the molecular basis of cell conversion(s) will certainly help to improve safety and reliability for regenerative medicine approaches in the future.

IN VITRO TRANSDIFFERENTIATION

Numerous studies have indicated that dedifferentiation and transdifferentiation during tissue regeneration can be triggered in vitro. Table 4.1 presents the most important cellular conversions and factors needed for transdifferentiation of specific cell types.

TABLE 4.1 Summary of In Vitro Transdifferentiation Protocols

Target Cells	Source Cells	Differentiation Factors	References
Cardiomyocytes	Mouse postnatal and cardiac dermal fibroblasts	Tbx5, Mef2c, Gata4, Mesp1	[54]
	Mouse embryonic fibroblasts	miR-1-1	[55]
	Mouse embryonic fibroblasts	Oct4, Sox2, Klf4, c-Myc	[56]
Neurons	Mouse embryonic fibroblasts	Ascl1, Brn2, Myt1l	[57]
	Human fetal and postnatal fibroblasts, human pluripotent stem cells	ASCL1, BRN2, MYT1L, NEUROD1	[58]
	Human primary dermal fibroblasts	miR-124, BRN2, MYT1L	[59]
	Human neonatal foreskin fibroblasts	miR-9/9*, miR-124 (+ ASCL1 and MYT1L as enhancers)	[60]
Dopaminergic neurons	Mouse embryonic fibroblasts and human adult fibroblasts	Ascl1, Nurr1, Lmx1a	[61]
	Human fibroblasts	Ascl1, Brn2, Myt1l, Lmx1a, FoxA2	[62]
Neural/ progenitor cells	Mouse fibroblasts	Oct4, Sox2, Klf4, and c-Myc (+ neural medium)	[63]
Motor neurons	Mouse and human fibroblasts	Ascl1, Brn2, Myt1l, Lhx3,Hb9, Isl1, Ngn2 (+ NEUROD1 for human)	[64]
Striatal neurons	Human postnatal and adult fibroblasts	miR-9/9*, miR-124 BCL11B, DLX1, DLX2, MYT1L	[65]
Hepatocytes	Mouse fibroblasts	Gata4, Hnf1α, Foxa3 (+ inactivation of $p19^{Arf}$)	[66]
	Mouse embryonic and adult fibroblasts	Hnf4α plus Foxa1 or Foxa2 or Foxa3	[47]

TABLE 4.1 Summary of In Vitro Transdifferentiation Protocols—cont'd

Target Cells	Source Cells	Differentiation Factors	References
Beta-cells	Mouse pancreatic exocrine cells	Ngn3, Pdx1, Mafa	[67]
Macrophages	Mouse embryonic fibroblasts	PU.1, C/EBPα	[68]
	Mouse B cells	C/EBPα, C/EBPβ	[69]
Photoreceptor cells	Human peripheral blood mononuclear cells	CRX	[70]
Corneal epithelial cells	Human fibroblasts	ΔNp63α, C/EBPδ, TCF4	[71]
Myocytes	Mouse embryonic fibroblasts	MyoD	[72]
Vascular smooth muscle cells	Mouse fibroblasts and neural crest stem cells	miR-145, miR-143	[73]

An important milestone in transdifferentiation is the successful conversion of fibroblasts to muscle cells using the single TF myogenic differentiation 1 (MyoD) [72]. In a similar study, ectopic eye structures are induced on the wings, legs, or the antenna of Drosophila by targeted expression of paired box protein pax-6 (pax6) [74]. These studies suggest that there are specific master regulators which are crucial to drive cell fate decisions. The most common strategy to identify key transcription factors for transdifferentiation is to simultaneously test a maximum number of known TFs and to gradually narrow to the most necessary ones. This approach is markedly limited by the costly and time-consuming experimental testing of sets of TFs. Recently, a bioinformatics tool called Mogrify has been used to predict TFs necessary to induce cell conversions [75]. It combines gene expression data with regulatory network information to find optimal TF combinations needed for transdifferentiation. The authors apply their model to 173 human cell types and 134 tissues and thus define an atlas of cellular transdifferentiation. Importantly, this system correctly predicts TFs used in previously described transdifferentiation protocols and predicts novel ones, which are subsequently validated. Such bioinformatics tools will undoubtedly be a valuable resource for further research on transdifferentiation and regenerative medicine.

The generation of functional neurons from somatic cells bypassing pluripotent cells is an active research area in neurobiology. Transdifferentiation into neurons is safer as it prevents neural outgrowth or unwanted proliferation of stem cells. This would be advantageous with respect to potential medical applications, and would allow autologous transplantation of neurons that are produced from the patient's own fibroblasts. An alternative to skin fibroblasts is the direct conversion of endogenous glial cells from inside the brain into functional neurons. Indeed, the generation of neurons from somatic cells is possible by forced expression of selected TFs. A screening approach showed that forced expression of specific TFs and/or miRNAs converts nonneuronal cells across germ layers into neuronal cells by direct lineage conversion [57–60].

Treating disease-specific phenotypes using transdifferentiated neurons requires a profound understanding of the complexity of neuronal development which is needed to for the efficient generation of functional neural subtypes. Toward this aim, transdifferentiated dopaminergic [61, 62] or spinal motor neurons [64] are generated using selected TFs that are known to play a role in the relevant neuronal specification during embryonic development. Generation of functional neural subtypes that can be successfully engrafted and integrated into the brain network can promote novel opportunities in drug screening assays and transplantation therapies.

FACTORS ENHANCING TRANSDIFFERENTIATION

Research on iPSC generation in the past decade has brought valuable answers regarding safety and approaches that enhance efficiency of cellular conversions. To reduce genomic integration of lenti- or retroviruses, reprogramming factors are introduced using nonintegrating Sendai viral vectors, episomal vectors, *piggyBac* transposons, purified recombinant reprogramming proteins or synthetic mRNAs. In transdifferentiation studies, the same strategies can be applied to reduce the risks of insertion-mutagenesis and oncogenic transformation. Unfortunately, many of the mentioned methods considerably reduce the already low reprogramming and transdifferentiation efficiency. To increase specificity and efficiency, numerous strategies are being explored. Cell culture under hypoxic conditions and addition of valproic acid and antioxidants such as ascorbic acid to the culture media are examples of factors that can enhance cellular conversions. It has also been shown that incorporating suitable scaffolds and matrixes for cell growth can facilitate transdifferentiation and reprogramming. As described by Downing et al. introducing microgrooves in cell-adhesive substrates mechano-modulates the cell's epigenetic profile and thus enhances cell conversion. The exact mechanisms underlying these approaches are not fully understood, but it is postulated that histone deacytelases and histone methyltrasferases play a key role in this process. Similar effects to those produced by microgrooves are also observed with nanofibrous scaffolds, suggesting that changes in cellular morphology may be responsible for modulation of their epigenetic state [76]. Recently, Gelmi et al. have presented an electroactive scaffold that provides electrical, mechanical, and topographical cues to cells differentiating into cardiomyocytes. Scaffolds coated with conductive polymer polypyrole are also capable of delivering direct electrical and mechanical stimulation to cells, mimicking the environment of heart tissue and enhancing conversion into cardiomyocytes [77].

Nowadays, much effort is being put in the identification of small molecules that can replace TFs during transdifferentiation and reprogramming [78]. It is known that single molecules can replace one or several TFs. One such example is the replacement of Sox2 by an inhibitor of the TGF-β receptor during somatic cell reprogramming [79]. However, the ultimate goal is to develop effective and safe transdifferentiation methods which are solely based on chemical compounds.

Promising results obtained by Cheng et al. demonstrate that functional neural progenitor cells (NPCs) can be induced directly from mouse embryonic fibroblasts, mouse tail-tip fibroblasts, and human urinary cells using a chemical cocktail, named VCR (V=VPA, an inhibitor of HDACs; C=CHIR99021, an inhibitor of GSK-3 kinases; and R=Repsox, an inhibitor of TGF-β pathway), under physiological hypoxic conditions. The derived cells resemble mouse brain derived NPCs with respect to their self-renewing abilities, gene expression profiles, and multipotent differentiation capacity. The authors also demonstrate that the small molecules originally used, could be successfully replaced with other inhibitors of histone deacetylation, glycogen synthase kinase, and TGF-beta pathways [80]. Similar studies on obtaining neural stem cells from fibroblasts are performed in normoxic conditions using A-83-01, CHIR99021, VPA, BIX01294, RG108, PD0325901, and vitamin C [81]. To convert human fibroblast directly into neurons and to bypass the NPC state, the cocktail proposed by Cheng et al. [80] can be additionally supplemented with forskolin, SP600125, GO6983, and Y-27632 [82]. Similar approaches are applied to obtain pancreatic cells. Pennarossa et al. converted human skin fibroblasts into pancreatic cells by exposing fibroblasts for 18h to the DNA methyltransferase inhibitor 5-azacytidine (AZA) followed by a three-step protocol (36 days) for the induction of endocrine pancreatic differentiation. The authors achieved an efficiency of ~35%, and demonstrated that the converted cells acquire an epithelial morphology and release insulin in response to physiological glucose challenge in vitro [83]. Insulin expressing cells are also generated using transdifferentiation of skin fibroblasts from type 1 diabetes patients in culture medium containing nicotinamide and exedin-4 [84]. In addition, Sasaki and coworkers converted a human hepatoma cell line (HepG2 cells) into pancreatic-like cells using CCl_4, D-galactosamine, and $ZnCl_2$ [85].

There are already some studies exemplifying how small molecules facilitate changes in cell fate decisions in vivo. Using *C. elegans*, Morgan et al. showed in 2010 that a small molecule converts sperm-producing cells into the female counterpart, the oocyte [86]. The authors tested small-molecules targeting the Ras-ERK signaling pathway and found that MEK inhibitor U0126-induced oocyte formation within 24h. Another in vivo study reports that cardiac fibroblasts can change into endothelial cells after cardiac injury by stimulation with the p53 activator "reactivation of p53 and induction of tumor apoptosis" (RITA) [87]. These two examples clearly demonstrate that chemically mediated switches between cell types can occur not only in vitro, but also in vivo [88].

Identifying drugs that are capable of inducing in vivo conversion of one cell type into another that in turn can be used to repair damaged tissue would undoubtedly revolutionize current regenerative medicine approaches. Hence, efforts on finding small molecules that possess high precision and selectivity in converting one somatic cell type to another need to be continued. Although the molecular mechanisms of small molecule-mediated transdifferentiation are to date not well understood, recent studies begin to shed light on the molecular basis and will hopefully contribute to a better understanding of their molecular basis [78, 88].

THERAPEUTIC TRANSDIFFERENTIATION

Proof-of-concept studies using animal models have been used to demonstrate that "therapeutic" in vivo transdifferentiation is possible for several disorders, including myocardial infarction (MI) and diabetes.

Despite significant advances in treatment over the past 10 years, the incidence of MI has become the leading cause of death worldwide. Each year 1.7 million Europeans die from MI and the annual health care cost associated with MI is around 60 billion Euro per year (European Cardiovascular Disease Statistics 2012). MI is caused by a blockage of a coronary artery with subsequent death of cardiomyocytes due to a lack of oxygen. The wound healing response results in the replacement of dead cardiac tissue by noncontractile scar tissue, leading to a loss of cardiac pumping capacity. Cardiac stem cell therapy can in theory replace the contractile units lost after MI because the stem cells can "reinforce" the scar and rebuild contractile muscle tissue. However, clinical trials of cardiac stem cell therapy have so far been of little success, and direct injection of stem cells has been shown to result in rapid death of these injected cells [89]. A valuable alternative to permit true cardiovascular regeneration and restoration of functional tissue would be to transdifferentiate scar-forming fibroblasts into cardiovascular cells. Recently, several studies provide proof that such an in vivo therapeutic transdifferentiation is indeed possible [54, 90]. In 2010, Ieda et al. showed that a combination of three TFs (Gata4, Mef2c, and Tbx5) can convert postnatal cardiac or dermal fibroblasts into spontaneously contracting cardiomyocyte cells in a rapid and effective way. Fibroblasts that were transduced with the three TFs and transplanted into the infarct zone of mice were able to differentiate into cardiomyocyte-like cells and restored heart function [54]. These preclinical studies will advance the understanding of cardiac transdifferentiation mechanisms and indicate that cells obtained in such way may have therapeutic value.

Another disorder that has been extensively studied to develop therapeutic transdifferentiation approaches is diabetes [91]. Type I diabetes is linked to reduced numbers of endocrine insulin-producing β-cells. Studies are focusing on possible restoration of these cells by coaxing other pancreatic cell types such as exocrine cells which are producing digestive enzymes [88]. The forced expression of three key TFs, Neurog3, Pdx1, and Mafa, in pancreatic exocrine cells has been shown to induce transdifferentiation into insulin-secreting β-cells. Interestingly, in this study adenovirus encoding the three TFs is directly injected into the pancreas [67] and showed that the microenvironment influences the outcome of in vivo transdifferentiation process. These data show that the glucose metabolism is influencing the efficacy of this process and that hyperglycemia, characterized by high glucose blood levels, is hindering the production of functional β-cells [92]. Another in vivo approach uses transdifferentiation of liver cells into insulin-producing β-cells. Ferber et al. demonstrated that adenoviral transduction of the TF pancreas/duodenal homeobox gene 1 (Pdx1), which is crucial for pancreatic development, resulted in insulin-producing cells in the liver and lowered blood glucose levels [93]. Liver cells can also be transformed into β-cells by using the three key TFs Neurog3, Pdx1, and Mafa [94]. Interestingly, it has been proven that transdifferentiated β-cells are not derived from hepatic cells, but from a cell population that expressed Sox9. These cells belong either to the bile duct epithelium or to hepatoblast-like progenitors. These results indicate that not all cells have transdifferentiation potential.

In vitro work has proven that human exocrine cells can be transdifferentiated into insulin-secreting β-cells [95]. In addition, β-cells can be derived from other human mature cell types such as fibroblasts [96]. However, procedures using nonpancreatic cells will likely depend on ex vivo transdifferentiation protocols since the cells are not directly targeting the proper in vivo niche in the pancreas. This might be a disadvantage for therapeutic approaches since cell conversions that can only be performed in vitro would require more invasive procedures including biopsies, in vitro transdifferentiation, and retransplantation into the patient.

The use of integrating viruses harbors the danger of mutagenesis and carcinogenesis. Therefore, nonintegrative approaches are needed to improve the safety of therapeutic in vivo approaches. Indeed, it has recently been shown that β-cell transdifferentiation can be achieved through transfection of synthetic mRNAs coding for the same three key TFs mentioned above (Neurog3, Pdx1, and Mafa) [97]. To date, this has only been described in vitro, but holds a promising approach for translation into preclinical in vivo studies. Interestingly, in the same study, Koblas and colleagues show that the transdifferentiation efficacy is enhanced by pretreatment with the DNA demethylating agent decitabine (also known as 5-Aza-2′-deoxycytidine), underlying the importance and relationships of chromatin accessibility, gene regulation, and cell differentiation (see also "Gene Regulation During Transdifferentiation" section).

CONCERNS AND FUTURE PERSPECTIVES ASSOCIATED WITH TRANSDIFFERENTIATION

As discussed, therapeutic in vivo transdifferentiation has been successfully applied in a variety of animal disease models. In order to bring these approaches closer to the clinics, it is necessary to understand and minimize the potential risks associated with in vivo transdifferentiation.

Although numerous examples of in vitro transdifferentiation hold promising therapeutic perspectives, it is likely that only some can be applied in vivo. The biggest advantage of transdifferentiation protocols over two-step protocols using iPSC generation and downstream differentiation is the missing intermediate step of pluripotency. iPSCs are thought to be more likely to develop tumors due to their unlimited proliferation capacity; however, further studies are needed to analyze and compare cells derived by transdifferentiation carefully to determine their epigenetic and genetic features. In addition, future efforts need to focus on identifying small compounds that can replace virally transduced factors to bring transdifferentiation closer to the clinics.

REFERENCES

[1] Shen CN, Burke ZD, Tosh D. Transdifferentiation, metaplasia and tissue regeneration. Organogenesis 2004;1(2):36–44. PubMed PMID: 19521559. Pubmed Central PMCID: PMC2633984.

[2] Dinsmore CE. A history of regeneration research. Milestones in the evolution of a science. Cambridge: Cambridge University Press; 1991.

[3] Colucci VL. Sulla rigenereazione parziale dell'occhio nei Tritoni-Istogenesi e sviluppo: studio sperimentale. Mem R Acad Sci Ist Bologna 1891;51:593–629.

[4] Wolff G. Entwicklungsphysiologische Studien. I. die Regeneration der Urodelenlinse. Wilhelm Roux Arch Entwickl Mech Org 1895;1:380–90.

[5] Jopling C, Boue S, Belmonte JCI. Dedifferentiation, transdifferentiation and reprogramming: three routes to regeneration. Nat Rev Mol Cell Biol 2011;12(2):79–89. PubMed PMID: WOS:000286465600012 [English].

[6] Sandoval-Guzman T, Wang H, Khattak S, Schuez M, Roensch K, Nacu E, et al. Fundamental differences in dedifferentiation and stem cell recruitment during skeletal muscle regeneration in two salamander species. Cell Stem Cell 2014;14(2):174–87. PubMed PMID: WOS:000330835800010 [English].

[7] Day RC, Beck CW. Transdifferentiation from cornea to lens in *Xenopus laevis* depends on BMP signalling and involves upregulation of Wnt signalling. BMC Dev Biol 2011;11:54. PubMed PMID: WOS:000295413000001 [English].

[8] Rink JC. Stem cell systems and regeneration in planaria. Dev Genes Evol 2013;223(1–2): 67–84. PubMed PMID: WOS:000313949700006 [English].

[9] Neuhof M, Levin M, Rechavi O. Vertically- and horizontally-transmitted memories—the fading boundaries between regeneration and inheritance in planaria. Biol Open 2016;5(9): 1177–88. PubMed PMID: WOS:000384077300001 [English].

[10] Adell T, Cebria F, Salo E. Gradients in planarian regeneration and homeostasis. Cold Spring Harb Perspect Biol 2010;2(1):a000505. PubMed PMID: WOS:000279880900003 [English].

[11] Wagner DE, Wang IE, Reddien PW. Clonogenic neoblasts are pluripotent adult stem cells that underlie planarian regeneration. Science 2011;332(6031):811–6. PubMed PMID: WOS:000290529900038 [English].

[12] Scimone ML, Kravarik KM, Lapan SW, Reddien PW. Neoblast specialization in regeneration of the planarian schmidtea mediterranea. Stem Cell Rep 2014;3(2):339–52. PubMed PMID: WOS:000340882400012 [English].

[13] Wenemoser D, Reddien PW. Planarian regeneration involves distinct stem cell responses to wounds and tissue absence. Dev Biol 2010;344(2):979–91. PubMed PMID: WOS:000280787100039 [English].

[14] Petersen CP, Reddien PW. Wnt signaling and the polarity of the primary body axis. Cell 2009;139(6):1056–68. PubMed PMID: WOS:000272622800009 [English].

[15] Petersen CP, Reddien PW. A wound-induced Wnt expression program controls planarian regeneration polarity. Proc Natl Acad Sci USA 2009;106(40):17061–6. PubMed PMID: WOS:000270537500037 [English].

[16] Beane WS, Morokuma J, Lemire JM, Levin M. Bioelectric signaling regulates head and organ size during planarian regeneration. Development 2013;140(2):313–22. PubMed PMID: WOS:000312741400007 [English].

[17] Poss KD, Wilson LG, Keating MT. Heart regeneration in zebrafish. Science 2002; 298 (5601):2188–90. PubMed PMID: 12481136.

[18] Jopling C, Sleep E, Raya M, Marti M, Raya A, Izpisua Belmonte JC. Zebrafish heart regeneration occurs by cardiomyocyte dedifferentiation and proliferation. Nature 2010; 464 (7288):606–9. PubMed PMID: 20336145. Pubmed Central PMCID: PMC2846535.

[19] Kikuchi K, Holdway JE, Werdich AA, Anderson RM, Fang Y, Egnaczyk GF, et al. Primary contribution to zebrafish heart regeneration by gata4(+) cardiomyocytes. Nature 2010; 464 (7288):601–5. PubMed PMID: 20336144. Pubmed Central PMCID: PMC3040215.

[20] Sleep E, Boue S, Jopling C, Raya M, Raya A, Izpisua Belmonte JC. Transcriptomics approach to investigate zebrafish heart regeneration. J Cardiovasc Med (Hagerstown) 2010; 11(5):369–80. PubMed PMID: 20179605.

[21] Knopf F, Hammond C, Chekuru A, Kurth T, Hans S, Weber CW, et al. Bone regenerates via dedifferentiation of osteoblasts in the zebrafish fin. Dev Cell 2011;20(5):713–24. PubMed PMID: WOS:000291141600016 [English].

[22] Jarriault S, Schwab Y, Greenwald I. A *Caenorhabditis elegans* model for epithelial-neuronal transdifferentiation. Proc Natl Acad Sci USA 2008; 105 (10):3790–5. PubMed PMID: 18308937. Pubmed Central PMCID: PMC2268801.

[23] Richard JP, Zuryn S, Fischer N, Pavet V, Vaucamps N, Jarriault S. Direct in vivo cellular reprogramming involves transition through discrete, non-pluripotent steps. Development 2011;138(8):1483–92. PubMed PMID: WOS:000288649400004 [English].

[24] Chen ZL, Yu WM, Strickland S. Peripheral regeneration. Annu Rev Neurosci 2007; 30:209–33. PubMed PMID: 17341159.

[25] Monje PV, Soto J, Bacallao K, Wood PM. Schwann cell dedifferentiation is independent of mitogenic signaling and uncoupled to proliferation role of cAMP and JNK in the maintenance of the differentiated state. J Biol Chem 2010;285(40):31024–36. PubMed PMID: WOS:000282135500067 [English].

[26] Yanger K, Zong Y, Maggs LR, Shapira SN, Maddipati R, Aiello NM, et al. Robust cellular reprogramming occurs spontaneously during liver regeneration. Genes Dev 2013; 27(7):719–24. PubMed PMID: 23520387. Pubmed Central PMCID: PMC3639413.

[27] Thorel F, Nepote V, Avril I, Kohno K, Desgraz R, Chera S, et al. Conversion of adult pancreatic alpha-cells to beta-cells after extreme beta-cell loss. Nature 2010;464(7292):1149–54. PubMed PMID: WOS:000276891100028 [English].

[28] Eguizabal C, Montserrat N, Veiga A, Izpisua Belmonte JC. Dedifferentiation, transdifferentiation, and reprogramming: future directions in regenerative medicine. Semin Reprod Med 2013; 31(1):82–94. PubMed PMID: 23329641.

[29] Slack JM. Homoeotic transformations in man: implications for the mechanism of embryonic development and for the organization of epithelia. J Theor Biol 1985; 114(3):463–90. PubMed PMID: 3927066.

[30] Merrell AJ, Stanger BZ. Adult cell plasticity in vivo: de-differentiation and transdifferentiation are back in style. Nat Rev Mol Cell Biol 2016; 17(7):413-25. PubMed PMID: 26979497.

[31] Tata PR, Mou HM, Pardo-Saganta A, Zhao R, Prabhu M, Law BM, et al. Dedifferentiation of committed epithelial cells into stem cells in vivo. Nature 2013;503(7475):218–23. PubMed PMID: WOS:000326894200042 [English].

[32] Jenuwein T, Allis CD. Translating the histone code. Science 2001; 293 (5532):1074-80. PubMed PMID: 11498575.

[33] Efroni S, Duttagupta R, Cheng J, Dehghani H, Hoeppner DJ, Dash C, et al. Global transcription in pluripotent embryonic stem cells. Cell Stem Cell 20082(5):437–47. PubMed PMID: 18462694. Pubmed Central PMCID: 2435228.

[34] Meshorer E, Misteli T. Chromatin in pluripotent embryonic stem cells and differentiation. Nat Rev Mol Cell Biol 2006; 7(7):540–6. PubMed PMID: 16723974.

[35] Voigt P, Tee WW, Reinberg D. A double take on bivalent promoters. Genes Dev 2013; 27 (12):1318–38. PubMed PMID: 23788621. Pubmed Central PMCID: 3701188.

[36] De Gobbi M, Garrick D, Lynch M, Vernimmen D, Hughes JR, Goardon N, et al. Generation of bivalent chromatin domains during cell fate decisions. Epigenetics Chromatin 2011; 4(1):9. PubMed PMID: 21645363. Pubmed Central PMCID: 3131236.

[37] Savic N, Bar D, Leone S, Frommel SC, Weber FA, Vollenweider E, et al. lncRNA maturation to initiate heterochromatin formation in the nucleolus is required for exit from pluripotency in ESCs. Cell Stem Cell 2014; 15(6):720–34. PubMed PMID: 25479748.

[38] Feldman N, Gerson A, Fang J, Li E, Zhang Y, Shinkai Y, et al. G9a-mediated irreversible epigenetic inactivation of Oct-3/4 during early embryogenesis. Nat Cell Biol 2006; 8(2): 188–94. PubMed PMID: 16415856.

[39] Epsztejn-Litman S, Feldman N, Abu-Remaileh M, Shufaro Y, Gerson A, Ueda J, et al. De novo DNA methylation promoted by G9a prevents reprogramming of embryonically silenced genes. Nat Struct Mol Biol 2008; 15 (11):1176–83. PubMed PMID: 18953337. Pubmed Central PMCID: 2581722.

[40] Sainsbury S, Bernecky C, Cramer P. Structural basis of transcription initiation by RNA polymerase II. Nat Rev Mol Cell Biol 2015; 16(3):129–43. PubMed PMID: 25693126.

[41] Spitz F, Furlong EE. Transcription factors: from enhancer binding to developmental control. Nat Rev Genet 2012; 13(9):613–26. PubMed PMID: 22868264.

[42] Iwafuchi-Doi M, Zaret KS. Cell fate control by pioneer transcription factors. Development 2016;143(11):1833–7. PubMed PMID: WOS:000389588800001 [English].

[43] Magnani L, Eeckhoute J, Lupien M. Pioneer factors: directing transcriptional regulators within the chromatin environment. Trends Genet 2011;27(11):465–74. PubMed PMID: WOS:000296485800004 [English].

[44] Taube JH, Allton K, Duncan SA, Shen LL, Barton MC. Foxa1 functions as a pioneer transcription factor at transposable elements to activate Afp during differentiation of embryonic stem cells. J Biol Chem 2010;285(21):16135–44. PubMed PMID: WOS:000277715900051 [English].

[45] Iwafuchi-Doi M, Zaret KS. Pioneer transcription factors in cell reprogramming. Genes Dev 2014;28(24):2679–92. PubMed PMID: WOS:000346335700001 [English].

[46] van Oevelen C, Kallin EM, Graf T. Transcription factor-induced enhancer modulations during cell fate conversions. Curr Opin Genet Dev 2013;23(5):562–7. PubMed PMID: WOS:000324970600010 [English].

[47] Sekiya S, Suzuki A. Direct conversion of mouse fibroblasts to hepatocyte-like cells by defined factors. Nature 2011;475(7356):390–3. 390-U148. PubMed PMID: WOS:000292911200045 [English].

[48] Takahashi K, Yamanaka S. Induction of pluripotent stem cells from mouse embryonic and adult fibroblast cultures by defined factors. Cell 2006 126(4):663–76. PubMed PMID: 16904174.

[49] Soufi A, Donahue G, Zaret KS. Facilitators and impediments of the pluripotency reprogramming factors' initial engagement with the genome. Cell 2012;151(5):994–1004. PubMed PMID: WOS:000311423500013 [English].

[50] Whyte WA, Orlando DA, Hnisz D, Abraham BJ, Lin CY, Kagey MH, et al. Master transcription factors and mediator establish super-enhancers at key cell identity genes. Cell 2013; 153(2):307–19. PubMed PMID: 23582322. Pubmed Central PMCID: 3653129.

[51] Lee J, Sayed N, Hunter A, Au KF, Wong WH, Mocarski ES, et al. Activation of innate immunity is required for efficient nuclear reprogramming. Cell 2012;151(3):547–58. PubMed PMID: WOS:000310529300012 [English].

[52] Ahrens TD, Lutz L, Lassmann S, Werner M. Turning skyscrapers into town houses: insights into Barrett's esophagus. Pathobiology 2017;84(2):87–98. PubMed PMID: 27656900.

[53] Colleypriest BJ, Ward SG, Tosh D. How does inflammation cause Barrett's metaplasia? Curr Opin Pharmacol 2009;9(6):721–6. PubMed PMID: WOS:000272445300008 [English].

[54] Ieda M, Fu JD, Delgado-Olguin P, Vedantham V, Hayashi Y, Bruneau BG, et al. Direct reprogramming of fibroblasts into functional cardiomyocytes by defined factors. Cell 2010;142(3):375–86. PubMed PMID: WOS:000280609100015 [English].

[55] Ivey KN, Muth A, Amold J, King FW, Yeh RF, Fish JE, et al. MicroRNA regulation of cell lineages in mouse and human embryonic stem cells. Cell Stem Cell 2008;2(3):219–29. PubMed PMID: WOS:000254017700009 [English].

[56] Efe JA, Hilcove S, Kim J, Zhou H, Ouyang K, Wang G, et al. Conversion of mouse fibroblasts into cardiomyocytes using a direct reprogramming strategy. Nat Cell Biol 2011;13(3):215–22. PubMed PMID: WOS:000287847600009 [English].

[57] Vierbuchen T, Ostermeier A, Pang ZP, Kokubu Y, Sudhof TC, Wernig M. Direct conversion of fibroblasts to functional neurons by defined factors. Nature 2010;463(7284):1035–41. PubMed PMID: WOS:000275108400027 [English].

[58] Pang ZPP, Yang N, Vierbuchen T, Ostermeier A, Fuentes DR, Yang TQ, et al. Induction of human neuronal cells by defined transcription factors. Nature 2011;476(7359):220–3. PubMed PMID: WOS:000293731900039 [English].

[59] Ambasudhan R, Talantova M, Coleman R, Yuan X, Zhu SY, Lipton SA, et al. Direct reprogramming of adult human fibroblasts to functional neurons under defined conditions. Cell Stem Cell 2011;9(2):113–8. PubMed PMID: WOS:000293990900008 [English].

[60] Yoo AS, Sun AX, Li L, Shcheglovitov A, Portmann T, Li YL, et al. MicroRNA-mediated conversion of human fibroblasts to neurons. Nature 2011;476(7359):228–31. PubMed PMID: WOS:000293731900041 [English].

[61] Caiazzo M, Dell'Anno MT, Dvoretskova E, Lazarevic D, Taverna S, Leo D, et al. Direct generation of functional dopaminergic neurons from mouse and human fibroblasts. Nature 2011;476(7359):224–7. PubMed PMID: WOS:000293731900040 [English].

[62] Pfisterer U, Kirkeby A, Torper O, Wood J, Nelander J, Dufour A, et al. Direct conversion of human fibroblasts to dopaminergic neurons. Proc Natl Acad Sci USA 2011;108(25):10343–8. PubMed PMID: WOS:000291857500063 [English].

[63] Kim J, Efe JA, Zhu SY, Talantova M, Yuan X, Wang SF, et al. Direct reprogramming of mouse fibroblasts to neural progenitors. Proc Natl Acad Sci USA 2011;108(19):7838–43. PubMed PMID: WOS:000290439500042 [English].
[64] Son Esther Y, Ichida Justin K, Wainger Brian J, Toma Jeremy S, Rafuse Victor F, Woolf Clifford J, et al. Conversion of mouse and human fibroblasts into functional spinal motor neurons. Cell Stem Cell. 2011 9/2/;9(3):205-18.
[65] Victor MB, Richner M, Hermanstyne TO, Ransdell JL, Sobieski C, Deng PY, et al. Generation of human striatal neurons by microRNA-dependent direct conversion of fibroblasts. Neuron 2014;84(2):311–23. PubMed PMID: WOS:000344167900010 [English].
[66] Huang PY, He ZY, Ji SY, Sun HW, Xiang D, Liu CC, et al. Induction of functional hepatocyte-like cells from mouse fibroblasts by defined factors. Nature 2011;475(7356):386–9. PubMed PMID: WOS:000292911200044 [English].
[67] Zhou Q, Brown J, Kanarek A, Rajagopal J, Melton DA. In vivo reprogramming of adult pancreatic exocrine cells to beta-cells. Nature 2008; 455 (7213):627–32. PubMed PMID: 18754011.
[68] Feng R, Desbordes SC, Xie HF, Tillo ES, Pixley F, Stanley ER, et al. PUA and C/EBP alpha/beta convert fibroblasts into macrophage-like cells. Proc Natl Acad Sci USA 2008;105(16):6057–62. PubMed PMID: WOS:000255356000021 [English].
[69] Xie HF, Ye M, Feng R, Graf T. Stepwise reprogramming of B cells into macrophages. Cell 2004;117(5):663–76. PubMed PMID: WOS:000221857900012 [English].
[70] Komuta Y, Ishii T, Kaneda M, Ueda Y, Miyamoto K, Toyoda M, et al. In vitro transdifferentiation of human peripheral blood mononuclear cells to photoreceptor-like cells. Biol Open 2016;5(6):709–19. PubMed PMID: WOS:000378109400005 [English].
[71] Cieslar-Pobuda A, Rafat M, Knoflach V, Skonieczna M, Hudecki A, Malecki A, et al. Human induced pluripotent stem cell differentiation and direct transdifferentiation into corneal epithelial-like cells. Oncotarget. 2016 Jun 2. 7(27):42314–42329 PubMed PMID: 27275539.
[72] Davis RL, Weintraub H, Lassar AB. Expression of a single transfected cDNA converts fibroblasts to myoblasts. Cell 1987; 51(6):987-1000. PubMed PMID: 3690668.
[73] Cordes KR, Sheehy NT, White MP, Berry EC, Morton SU, Muth AN, et al. miR-145 and miR-143 regulate smooth muscle cell fate and plasticity. Nature 2009;460(7256):705–10. PubMed PMID: WOS:000268670300031 [English].
[74] Halder G, Callaerts P, Gehring W. Induction of ectopic eyes by targeted expression of the eyeless gene in Drosophila. Science 1995;267(5205):1788–92.
[75] Rackham OJ, Firas J, Fang H, Oates ME, Holmes ML, Knaupp AS, et al. A predictive computational framework for direct reprogramming between human cell types. Nat Genet 2016; 48(3):331–5. PubMed PMID: 26780608.
[76] Downing TL, Soto J, Morez C, Houssin T, Fritz A, Yuan F, et al. Biophysical regulation of epigenetic state and cell reprogramming. Nat Mater 2013; 12 (12):1154–62. PubMed PMID: 24141451.
[77] Gelmi A, Cieslar-Pobuda A, de Muinck E, Los M, Rafat M, Jager EW. Direct mechanical stimulation of stem cells: a beating electromechanically active scaffold for cardiac tissue engineering. Adv Healthc Mater 2016; 5(12):1471–80. PubMed PMID: 27126086.
[78] Xu A, Cheng L. Chemical transdifferentiation: closer to regenerative medicine. Front Med 2016; 10(2):152–65. PubMed PMID: 27142989.
[79] Ichida JK, Blanchard J, Lam K, Son EY, Chung JE, Egli D, et al. A small-molecule inhibitor of tgf-Beta signaling replaces sox2 in reprogramming by inducing nanog. Cell Stem Cell 2009 5(5):491–503. PubMed PMID: 19818703. Pubmed Central PMCID: 3335195.

[80] Cheng L, Hu W, Qiu B, Zhao J, Yu Y, Guan W, et al. Generation of neural progenitor cells by chemical cocktails and hypoxia. Cell Res 2014; 24(6):665–79. PubMed PMID: 24638034. Pubmed Central PMCID: 4042166.

[81] Han YC, Lim Y, Duffieldl MD, Li H, Liu J, Abdul Manaph NP, et al. Direct reprogramming of mouse fibroblasts to neural stem cells by small molecules. Stem Cells Int 2016; 2016:4304916. PubMed PMID: 26788068. Pubmed Central PMCID: 4695670.

[82] Hu W, Qiu B, Guan W, Wang Q, Wang M, Li W, et al. Direct conversion of normal and Alzheimer's disease human fibroblasts into neuronal cells by small molecules. Cell Stem Cell 2015; 17(2):204–12. PubMed PMID: 26253202.

[83] Pennarossa G, Maffei S, Campagnol M, Tarantini L, Gandolfi F, Brevini TA. Brief demethylation step allows the conversion of adult human skin fibroblasts into insulin-secreting cells. Proc Natl Acad Sci USA 2013; 110 (22):8948–53. PubMed PMID: 23696663. Pubmed Central PMCID: 3670366.

[84] Pereyra-Bonnet F, Gimeno ML, Argumedo NR, Ielpi M, Cardozo JA, Gimenez CA, et al. Skin fibroblasts from patients with type 1 diabetes (T1D) can be chemically transdifferentiated into insulin-expressing clusters: a transgene-free approach. PLoS One 2014; 9(6):e100369. PubMed PMID: 24963634. Pubmed Central PMCID: 4070975.

[85] Kanoh Y, Tomotsune D, Shirasawa S, Yoshie S, Ichikawa H, Yokoyama T, et al. In vitro transdifferentiation of HepG2 cells to pancreatic-like cells by CCl(4), D-galactosamine, and ZnCl(2). Pancreas 2011; 40(8):1245–52. PubMed PMID: 21989025.

[86] Morgan CT, Lee MH, Kimble J. Chemical reprogramming of *Caenorhabditis elegans* germ cell fate. Nat Chem Biol 2010; 6(2):102–4. PubMed PMID: 20081824. Pubmed Central PMCID: PMC2808631.

[87] Ubil E, Duan J, Pillai ICL, Rosa-Garrido M, Wu Y, Bargiacchi F, et al. Mesenchymal-endothelial transition contributes to cardiac neovascularization. Nature 2014;514(7524): 585–90. PubMed PMID: WOS:000343801500039 [English].

[88] Wagner BK. Grand challenge commentary: chemical transdifferentiation and regenerative medicine. Nat Chem Biol 2010; 6(12):877–9. PubMed PMID: 21079596.

[89] Terrovitis JV, Smith RR, Marban E. Assessment and optimization of cell engraftment after transplantation into the heart. Circ Res 2010; 106(3):479–94. PubMed PMID: 20167944. Pubmed Central PMCID: 2826722.

[90] Sayed N, Wong WT, Cooke JP. Therapeutic transdifferentiation: can we generate cardiac tissue rather than scar after myocardial injury? Methodist Debakey Cardiovasc J 2013; 9(4):210–2. PubMed PMID: 24298312. Pubmed Central PMCID: 3846075.

[91] Cavelti-Weder C, Li W, Zumsteg A, Stemann M, Yamada T, Bonner-Weir S, et al. Direct reprogramming for pancreatic Beta-cells using key developmental genes. Curr Pathobiol Rep 2015; 3(1):57–65. PubMed PMID: 26998407. Pubmed Central PMCID: PMC4797339.

[92] Cavelti-Weder C, Li W, Zumsteg A, Stemann-Andersen M, Zhang Y, Yamada T, et al. Hyperglycaemia attenuates in vivo reprogramming of pancreatic exocrine cells to beta cells in mice. Diabetologia 2016; 59(3):522–32. PubMed PMID: 26693711. Pubmed Central PMCID: PMC4744133.

[93] Ferber S, Halkin A, Cohen H, Ber I, Einav Y, Goldberg I, et al. Pancreatic and duodenal homeobox gene 1 induces expression of insulin genes in liver and ameliorates streptozotocin-induced hyperglycemia. Nat Med 2000; 6(5):568–72. PubMed PMID: 10802714.

[94] Banga A, Akinci E, Greder LV, Dutton JR, Slack JM. In vivo reprogramming of Sox9+ cells in the liver to insulin-secreting ducts. Proc Natl Acad Sci USA 2012109 (38):15336–41. PubMed PMID: 22949652. Pubmed Central PMCID: PMC3458366.

[95] Lemper M, Leuckx G, Heremans Y, German MS, Heimberg H, Bouwens L, et al. Reprogramming of human pancreatic exocrine cells to beta-like cells. Cell Death Differ 2015; 22(7):1117–30. PubMed PMID: 25476775. Pubmed Central PMCID: PMC4572860.
[96] Zhu S, Russ HA, Wang X, Zhang M, Ma T, Xu T, et al. Human pancreatic beta-like cells converted from fibroblasts. Nat Commun 2016; 7:10080. PubMed PMID: 26733021. Pubmed Central PMCID: PMC4729817.
[97] Koblas T, Leontovyc I, Loukotova S, Kosinova L, Saudek F. Reprogramming of pancreatic exocrine cells AR42J into insulin-producing cells using mRNAs for Pdx1, Ngn3, and MafA transcription factors. Mol Ther Nucleic Acids 2016; 5:e320. PubMed PMID: 27187823. Pubmed Central PMCID: PMC5014516.

GLOSSARY

Chromatin Chromatin is the complex of DNA and its bound proteins, like histones. It is the core structure of gene expression control. Epigenetic features like histone acetylation or DNA methylation, forms euchromatin, associated with active gene transcription, or heterochromatin, which represses gene expression.

Dedifferentiation Dedifferentiation describes that a mature cell goes back into a progenitor state during "natural" occurring processes. This can be in part also observed during tumorigenesis when mature cells gain stem cell-like features during oncogenic transformation.

Reprogramming The technique of reprogramming converts mature, differentiation cells into stem cells. This is induced by forced expression of distinct transcription factors. The original work of Yamanaka is based on four transcription factors (Oct3/4, Sox2, c-Myc, and Klf4), which is still the most common approach. The process can be enhanced or also induced by small molecules.

Transcription factor Transcriptions factors are proteins involved in gene regulation, either by promoting active gene transcription or by gene repression. Transcription factors act in complex networks with other transcription factors or chromatin remodeling complexes. A special subset of transcription factors ("pioneer factors") can bind to closed chromatin and start chromatin remodeling.

Transdifferentiation Transdifferentiation, also called direct reprogramming, describes the conversion of one cell type into another without going through the pluripotent cell state.

Chapter 5

Generation of Organs Based on Decellularized Extracellular Matrix Scaffolds

Shiva Akbari-Birgani[*,†], Maryam Tahmasebi Birgani[‡,§], Hossein Ansari[‡]
[*]*Faculty of Biological Sciences, Institute for Advanced Studies in Basic Sciences (IASBS), Zanjan, Iran*
[†]*Center for Research in Basic Sciences and Contemporary Technologies, Institute for Advanced Studies in Basic Sciences (IASBS), Zanjan, Iran*
[‡]*Department of Medical Genetics, School of Medicine, Ahvaz Jundishapur University of Medical Sciences, Ahvaz, Iran*
[§]*Cellular and Molecular Research Center, Ahvaz Jundishapur University of Medical Sciences, Ahvaz, Iran*

INTRODUCTION

One of the biggest problems in today's medicine is the lack of donated organs for patients who suffer from organ failure and therefore need to receive an organ transplant. This is outstanding especially in case of spinal cords injuries and heart or kidney failure patients [1–3]. Although the number of such patients is increasing year by year, the supply of donated organs is severely restricted. Such limitation also can be influenced firstly by the immune incompatibility between donors and recipients to avoid the transplant rejection and then by the limitation of organ storage or even satisfaction of donor's family to organ donation [4]. Therefore, needs to develop the alternative strategies seem urgent. In 1999, William Haseltine coined the term "regenerative medicine" as a novel method of therapy where the tissue engineering, stem cell (SC) biology, and nuclear transfer comes together to replace an injured tissue [5]. He found that embryonic SCs capable to differentiate into the all human body cell types [6]. Although William's definition may seem simplistic at first, by looking deeper into the term, we will find how regenerative medicine can be promising in near future and make dramatic changes in the treatment of patients whose organs are seriously injured and failed. Today, regenerative medicine has allocated a special place to itself and different therapeutic protocols have been developed although most of them are at the beginning of the route.

Stem Cells and Biomaterials for Regenerative Medicine. https://doi.org/10.1016/B978-0-12-812258-7.00005-8

Cell therapy, autologous transplantation, growth factor-mediated therapy, recombinant protein therapy, tissue engineering, and gene therapy are some examples from advances in the field of regenerative medicine. In this chapter, we are going to focus on the transplantable organ regeneration strategy and explain how the today's medicine could benefit from such protocols [7–10].

To date, organ transplantation is the only known definitive therapeutic strategy for patients suffering from end-stage organ failure. It faces with serious issues such as the immune response to the transplanted organ, which causes transplantation rejection, and donor organ shortage. These issues have led to the complication of this therapeutic approach [11, 12]. These challenges led the researchers to use the synthetic materials as an alternative source for transplantation. For the first time, in 2011, an artificial trachea was constructed for a 36-year-old patient and transplanted successfully. The synthetic trachea was made of the patient's SCs, which were embedded in a plastic replica of the patient's actual trachea and incubated in the presence of growth factors [13]. This finding established when the affected organs or tissues are replaced with one reconstructed by the patient-derived cells, the best result can be achieved. Therefore, organ regeneration lightened up a new way to treat injured organs or tissues and nowadays, due to the progress and innovation in the field of bioengineering, it has been a promising strategy to construct the alternative organs that replace injured organs. Recently, successes have achieved in small animal models using decellularized whole-organ scaffolds. As illustrated in Fig. 5.1, in this approach, briefly, decellularized scaffolds are generated using physical, chemical, and enzymatic procedures, after that the host cells are incorporated to a prepared scaffold, so the cells are supported by the engineered scaffolds in a suitable culture environment, and finally it is transplanted to the host [14, 15]. Three-dimensional (3D) bioprinting, another outstanding achievement in the current era, is a powerful tool to build a desired cell-laden construct. It is notable that there are still many challenges such as regeneration of complex organ (e.g., heart, lungs, kidney, liver, and pancreas) as well as the several technological obstacles. However, there are many complications that need to be addressed, generation of transplantable organs holds promise for organ replacement and remains of great interest for scientists in the field of regenerative medicine.

COMMON TYPES OF SCAFFOLDS

Today, tissue engineering uses various scaffolds to make tissue and organs. Normally, three groups of biomaterials are used in the fabrication of scaffolds: ceramics (e.g., Hydroxyapatite (HA) and Tricalcium phosphate (TCP)), synthetic polymers (e.g., polystyrene, poly-L-lactic acid (PLLA), and polyglycolic acid (PGA) poly-DL-lactic-*co*-glycolic acid (PLGA)), natural polymers (e.g., collagen, various proteoglycans, alginate-based substrates, and chitosan). Each group of materials presents its own strengths and weaknesses. These features are the basis for choosing a material to reconstruct a particular tissue. For ex-

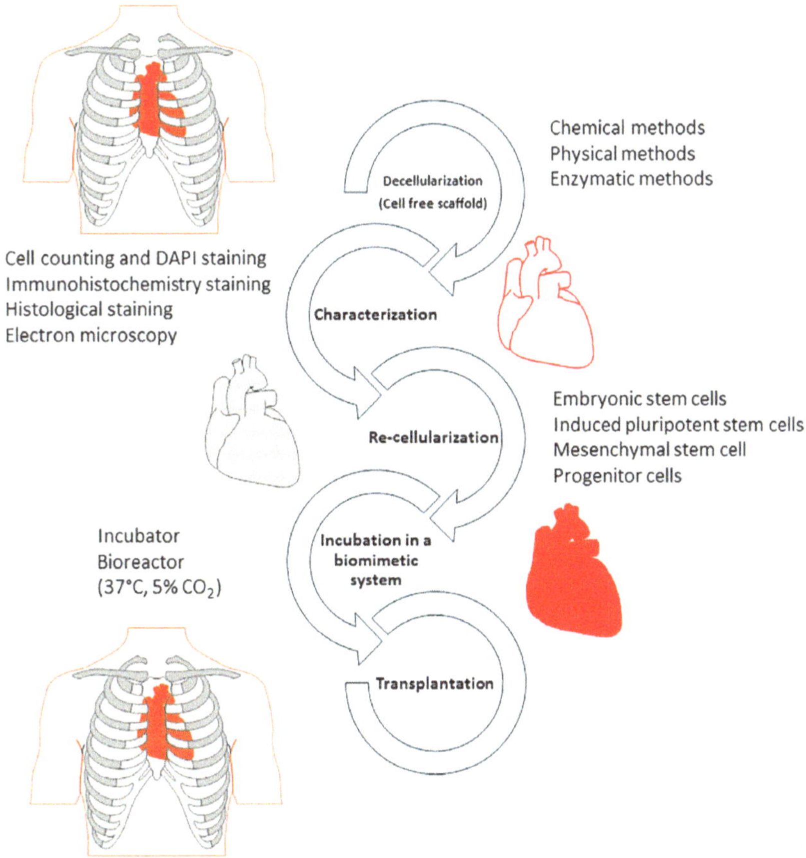

FIG. 5.1 A schematic representation of the overall steps related to the organ generation.

ample, ceramics due to their high mechanical stiffness are a good choice to regenerate bone and dental orthopedic surgery. In contrast, natural polymers are not suitable candidates for orthopedic purposes due to their mechanical weakness. In order to overcome these problems, a new generation of biomaterials emerged, named as combined materials. In fact, combined materials are a mixture of ceramics into polymer-based scaffolds or a complex of synthetic polymers with natural polymers (reviewed in Refs. [16–18]).

Fabricating of scaffolds from different sources requires attention to some of the critical concerns:

- *Biocompatibility*: The scaffolds should not stimulate the immune response.
- *Biodegradability*: The transplanted scaffolds required be gradually degraded, removed, and replaced with the newly synthesized ones in the body. On the other hand, degradation of the transplanted scaffolds should not produce toxic or interfering compounds that damage to the other part of the body.

- *Mechanical properties*: The mechanical properties such as stiffness, elasticity, tensile load of the grafting scaffold necessitate corresponding to the desired texture.
- *Scaffold architecture*: The structure of scaffold should be porous so that the cells penetrate to the scaffold and the cellular waste can be removed.

According to the concerns mentioned, the other suitable candidate is the intact extracellular matrix (ECM) of mammalian tissue. Nowadays, ECMs have broadly been implemented as the tissue scaffolds and applied in preclinical studies and clinical applications [19]. ECM is a complex network of matrix substances, functional proteins, proteoglycans, and signaling molecules, which naturally are produced by resident cells of a tissue. This matrix is considered as a means of the cell-cell interaction and cell interaction with the environment around it. In each tissue, cells align to a specific three-dimensional (3D) organization. In addition to the structural role of ECM in tissue formation, ECM remodeling plays a pivotal role in tissue hemostasis and development and undergoes modification in disease such as cancer and fibrosis [20–22]. Biomechanical and viscoelastic properties of ECM and its potential for host cell attachment enable it for tissue transplantation and remodeling. ECM strongly affects the cellular behavior and differentiation. The biomechanical features of ECM, such as strength and stiffness, are determined by the structure of the collagen fibers [23]. Depending on the type of tissue, 3D structure, function, and composition of ECM vary. However, some components such as collagen, laminin, fibronectin, and hyaluronic acid are general. Therefore in tissue grafting, 3D organization and components of the selected ECM should be matched with target tissue to mimic the chemical and mechanical properties of the actual tissue or organ, well [24]. Moreover, ECMs are capable to be modified. The modified ECMs are still biodegradable unless irreversible crosslinks occur between the molecules during processing [25, 26].

CONSTRUCTION OF CELL-FREE SCAFFOLDS

Decellularization

Cell-free scaffolds are derived from tissue decellularization methods. Tissue decellularization is referred to a set of physical, chemical, and enzymatic methods that removes cells from the tissues while preserving the natural content, tissue architecture, mechanical properties, and biological function of that tissue. Today, due to advances in both the knowledge and the instruments related to this method, whole organ decellularization is also possible. Indeed, construction of whole organ cell-free scaffolds is one of the brightest achievements in recent years.

Various types of methods are used to separate cells from the tissue, the most common of which are the following.

Decellularization by Perfusion Through the Vasculature

In this method, decellularizing reagents such as chemical and biological agents are used. Some of the commonly used decellularizing agents have been listed in Table 5.1 (an overview of tissue and whole organ decellularization processes). These reagents are passed through the vessels under physiological pressure. Moreover, airways in the lungs and bile ducts in the liver and the ureter in the kidney are the alternative ways to pass the reagents [27]. Indeed, the perfusion is an efficient strategy to deliver decellularizing agents to the cells; therefore, it is a preferential method to perform tissue and whole organ decellularization processes, and obtain whole-organ scaffolds [28]. Remarkably, it has been successfully applied to whole kidney, heart, and lung [29–33].

Immersion and Agitation

This method is used to decellularize the tissues that do not have access to the vascular system. In this method, the tissue is first immersed in a decellularizing solution, and it is simultaneously agitated with reagents to penetrate into the tissue. In this case, the cellular density and tissue thickness are very important for determining the duration of proximity. Examples of tissues that have been decellularized in this way include skeletal muscle and skin [34, 35].

The two methods mentioned are the most commonly used strategies to date. In some cases, other methods have been applied, for example, Sierad et al. developed a pressure gradient for cell removal of aortic roots and they achieved functional heart valve scaffolds [36]. Supercritical fluid is another alternative for cell removal, which has been reported for decellularization of aorta and cornea [37, 38]. Other approaches introduced by other researchers include thermal shock or freeze-thaw cycle disruption, hydrostatic pressure, and ultrasound [27, 39, 40]. It should be mentioned since each route has its advantages and disadvantages, the combination of physical, chemical, and enzymatic methods may be used to decellularize tissues and develop the cell-free ECM scaffolds. For example, in a protocol for decellularizing entheses provided by Xu et al.,

TABLE 5.1 List of Commonly Used Decellularizing Agents

Chemical Agents	Biologic Agents
Acids and bases Hypotonic and hypertonic solutions Nonionic detergents; Triton X-100 Ionic detergents; sodium dodecyl sulfate (SDS) sodium deoxycholate Triton X-200 Zwitterionic detergents; CHAPS sulfobetaine-10 and -16 (SB-10, SB-16) Solvents; alcohols acetone tributyl phosphate (TBP)	Enzymes; nucleases, trypsin, dispase Chelating agents (EDTA, EGTA)

a combination of chemical and physical methods, including freeze-thaw cycles, ultrasound, perfusion, and hydrostatic washing were used to achieve a highly efficient decellularization protocol [41]. In other protocol provided by Aubin et al., a whole-heart decellularization was performed using a software-controlled automated coronary perfusion with standard decellularization detergents [32].

Characterization of Cell-Free ECM Scaffolds

Decellularization is not the final step in the preparation of cell-free scaffolds because the decellularized scaffolds need to be characterized using various methods to confirm that there are no residual cell or alteration in ECM architecture. Cell remnants such as DNA can trigger the immune response of host in next steps of transplantation. In addition, studies show that changes in matrix architecture and residual detergent components not only can inhibit cell proliferation but also led to the cytotoxic effects [42–44]. Hence, ECM characterization is an inevitable step in the construction of the cell-free scaffolds.

Commonly used techniques to characterize the obtained scaffolds include histological stains to detect cellular and nuclear remnants such as hematoxylin and eosin (H&E) and 4′,6-diamidino-2-phenylindole (DAPI) staining, quantification of residual DNA and nuclear fragments using Quant-iT PicoGreen assay, immunohistochemistry to identify the composition of acellular matrix, electron microscopy to visualize ECM architecture; mechanical testing by the ball-burst test, a measure of strength in response to multiaxial loading [45]. In some studies, researchers in addition to the characterization of the prepared cell-free ECM, performed a cytotoxicity test to exclude the presence of soluble toxins in the biomaterial [31].

In a study by Chi Ting Au-Yeung et al., biophysical features of porcine cardiac ECM, after recellularization with human MSCs, were investigated. Their finding resulted in the identification of cells as the "biological plasticizers because the results showed that recellularization of acellular scaffold would restore some of the original biophysical properties of the myocardial tissue and make it more stabilized" [46].

Recellularization of Cell-Free ECM Scaffolds

Another challenging stage in organ regeneration is repopulation of an acellular ECM with cells (e.g., SCs or progenitor cells) and inculcation under physiological-like conditions in bioreactor systems. The cellular sources used in this strategy are diverse, including SCs, induced pluripotent SCs, adult SCs; we provide more details in next section. This stage also has a special sensitivity in such a way that a poor decellularization or recellularization may result in a failure of organ transplantation [42–44]. Usually for recellularization, a single cell suspension is injected into the pre-existing ECM scaffold and then depending on the cell type, the seeded scaffold is transferred to a tube containing the specific culture medium and incubated in a 37°C, 5% CO incubator, in static or bioreactor conditions, for a period of time. During incubation, media need to be changed. Culture medium and its ingredients vary by the cell type and the tissue of interest. For instance, in a work by Song et al., in order to accelerate in vitro nephrogenesis and maturation of NKCs in acellular kidney scaffold, in vivo maturation signals such as glucocorticoids and catecholamines were added to the culture media [47]. When bioreactors are used, preincubation in static condition is recommended, because it will provide enough time for adhesion of cells to the scaffold so it can increase the recellularization efficiency [48]. The number of injected cells and the duration of incubation vary by organ type and size and require to be optimized.

Three-Dimensional Bioprinting Strategy

One of the brilliant achievements in the field of tissue engineering is the invention of three-dimensional bioprinting methods. Using 3D bioprinting, instead of manipulating the primary tissue and decellularization, scaffolds are constructed by layer-by-layer printing of the biomaterial and simultaneous cell printing into tissue constructs with the predetermined architectures. Inkjet bioprinting, micro-extrusion systems, acoustic picoliters, laser-assisted bioprinting, stereolithography are various technologies have been applied to do bioprinting and fabricate the cell-laden constructs (reviewed in Ref. [49]). Recently, researchers have achieved a success in developing the cardiac patches using biomaterial-free 3D bioprinting. In this report, hiPSCs were used as a cell source and differentiated to functional cardiomyocytes. Transplantation of the fabricated cardiac patches to rat myocardium has showed its vascularization ability, marking a great success in tissue engineering methodologies [50]. In other report, researchers succeeded to build skin using keratinocytes and fibroblasts as the cellular source. The skin reconstruction has wide applications in transdermal and topical formulation discovery, dermal toxicity studies, and in designing autologous grafts for wound healing [51]. A novel bioprinting system with the capability to direct human cartilage repair has been developed. Poly(ethylene glycol) dimethacrylate with human chondrocytes were printed to repair defects in osteochondral plugs (3D biopaper) in layer-by-layer assembly [52]. The potential of 3D bioprinting technology for application in tooth regeneration has been evaluated by using human dental pulp cells (hDPCs) mixture as bioink [53]. In a work by Shim et al., a multilayered 3D construct was prepared using a mixture of human mesenchymal stromal cells with a hydrogel and applied to the knee joints of rabbits. The reported construct showed the regenerative ability for the reconstruction of an osteochondral tissue in the animal model [54].

Cell Sources for Organ Regeneration

Stem Cells

Following the discovery of SCs, it became clear that these cells could be considered as a material for cell therapy. In fact, the main factor behind the formation of such idea was the presence of three fundamental characteristics in these cells: self-renewal, clonogenicity, and plasticity, which present the capacity of proliferation, producing the identical daughter cells (symmetric divisions), and differentiation into the other cells, respectively (consequence of asymmetric divisions) [55]. Regarding the differentiation capacity, SCs are classified to the pluripotent, multipotent, and oligopotent in which the pluripotent are SCs with the ability to differentiate into all three embryonic germ layers: mesoderm-, endoderm-, and ectoderm-derived lineages so are in interest. Embryonic SCs are one of the good examples of pluripotent SCs although they produce teratoma in mice model [56]. Of course, isolation of SCs was considered as a milestone in cell therapy, which started at 1999 and has been continued until now. In cell therapy, we seek to restore the function of a damaged tissue to its normal state. But that's not just a cell-based infusion and requires a suitable scaffold to retain the cells in the target tissue and of course, guide them toward differentiating into the appropriate cells.

Induced-Pluripotent Stem Cells

At 2006, Takashi and Yamanaka showed that introducing the genes *KLf-4*, *Oct-4*, *Sox-2,* and *C-myc* into the fibroblast cells can induce the pluripotent state in these cells. The resulted cells express the surface marker like those expressing on embryonic SCs [57]. Similar experiments were performed afterward, which was termed as induced pluripotent stem cells (IPSC) technology or reprogramming. However, the oncogenic potential of some of these genes and their transferring vector are the disadvantages connected to iPSC [58]. These technologies have been discussed to greater extent in previous chapters.

Fetal-Derived SCs as a Modern Resource of SC

Use of embryonic cells has been recently considered by researchers due to some aspects. The cells can be isolated from fetal tissues like blood and bone marrow or from extraembryonic structures including amniotic fluid (AF), umbilical cord blood, Wharton's jelly, and placenta [59, 60]. These sources are promising as the procedures of their isolation are less invasive. Besides, the ethical issues are more subtle than the previous mentioned methods and harvesting of a large amount of tissue without ethical limitation is allowed [61]. In the following section, we explain components of fetal tissue as novel cell supply in regenerative medicine.

Amniotic Fluid Stem Cells

The amniotic fluid stem cells (AFSC) were firstly isolated by De Coppi as cells that express both embryonic and adult markers [62]. They exhibit strong proliferative potential and are capable to divide every 25–38h [61]. The AFSCs can be easily obtained through amniocentesis, that is typically conducted in the course of a prenatal diagnostic test to analyze the chromosomal abnormality or some types of birth defects (depending on the indication of the test) in a fetus of a pregnant mother at 14–16 weeks of gestation [63]. The AFSC can be easily isolated from low volume of amniotic fluid without any embryo manipulation (10^8 cells from 5 mL of AF). They have the low risk of immunogenicity and infection as well as high compatibility rate. In spite of embryonic SCs, AFSC is not tumorigenic in vivo and shows a stable karyotype even after a long passage in culture [61]. The latter observation was obtained through the study of the expression of SC marker such as Oct4, Nanog, Sox-2, and SSEA even after 18 subcultures [64]. Regarding the differentiation capacity, AFSC are multipotent. They have the plasticity to differentiate to adipocyte, osteoblast, myoblast, chondrocyte, neuron, and some other differentiated cells under a specific condition. It has been demonstrated that such fetal-derived SCs have more proliferation and differentiation capacity than the adult stem-cell-like mesenchymal SCs [65]. In spite of such advantages, some questions still exist about AFSC; first of all, can AFSCs keep their undifferentiated state following the graft to the target site to produce a long therapeutic effect? And second, do these fetal cells induce the DNA damage and stop the cell cycle event? [61] These are an important issue that should be addressed in future researches.

Placenta-Derived SCs

Although the clinical importance of mesenchymal SCs has been confirmed in pieces of literature, there are some disadvantages connected to these cells such as invasive and painful sampling from bone marrow (the typical source of MSCs) as well as low yield of isolation [66]. Additionally, it has demonstrated that the stemness properties of bone marrow SCs decrease with age [66]. Therefore, there is need to find an alternative to cope with such obstacles. Placenta-derived fetal tissue like chronic villi (CV) is now in interest due to several advantages [67]. Trophoblast has been originated from the outer cell mass of morula in early weeks of gestation and eventually forms the placenta during the human development. In fact, placenta acts as a bridge to attach the fetus to the uterine wall. Besides, the placenta uptakes the nutrient and eliminates the waste material from mother's blood. It also provides pregnancy's hormones [68, 69]. Mature placenta (third trimester of gestation) is composed of two fetal and maternal portions. The fetal portion originated from CV, smooth chorion, chorion plate, and amnion [68, 70]. The first-trimester placenta seems to be a novel supply for SCs as the placenta is only composed from fetal portion in this stage. It has shown that first trimester chronic villus (FTCV)-originated cells are the novel fetal supply for MSC extraction. Growing body of evidence demonstrated that CV-originated MSCs have a high rate of proliferation, genomic stability, and plasticity higher than the bone-marrow-derived ones [68]. They express the master genes of pluripotency *Sox-2* and *Nanog* as markers of stemness even in long culturing [71]. It has also found that CV-MSCs did not form teratoma following in vivo transplantation into the immunodeficient mice. The cells are also less immunogenic than the other sources of MSCs, which makes them suitable for transplantation [72]. Similar to the AFSCs, CV SCs can be obtained through chronic villous sampling procedure at 11 weeks of gestation or even at birth. This is also the reason that CV-SCs have been considered as novel candidates in regenerative medicine or treating of congenital birth defects [73].

Umbilical Cord Blood SCs

Fetus's umbilical cord (UC) develops from the placenta and contains the placenta remaining blood after baby birth. Although the UC is clinically considered as waste, today's science has shown that cord blood is enriched with SCs (like MSC) and hematopoietic SCs, which can be considered in regenerative medicine. In fact, UC protects umbilical vessels from torsion [74]. UC contains two arteries and one vein, which are covered with glycoprotein-enriched mucus called Wharton's jelly. This jelly structure is covered by amniotic epithelium and included with MSCs [75]. UC-derived SCs could be noninvasively collected after the birth of a baby and the resulting cells are multipotent [74]. UC-derived SCs can differentiate into the adipocytes, osteoblasts, chondrocytes, neurons, and hepatocytes [76]. The UC-MSCs are now in interest for autologous and allogeneic purposes. The only problem connected to the UC-MSCs is the fetus health, so tracking of the chromosomal stability should be considered [74].

EXAMPLES OF GENERATED ORGANS BASED ON REPRODUCTION OF CELL-FREE ECM SCAFFOLDS

In recent years, several successful examples of bioengineered organs have been developed based on the reproduction of cell-free ECM scaffolds. In next section, we briefly describe some of them.

Bioengineered Lung

Daly et al. investigated the reconstruction of a whole mouse lung through the incorporation of the bone-marrow-derived mesenchymal stromal cells (MSCs) to the fully characterized of decellularized whole mouse lungs. They incubated the cell-seeded constructs in basal MSC media and in small airways growth media (SAGM), over a period of time. They showed that decellularized lung maintained structural features of native lungs and generate a functional lung tissue ex vivo that can be used for surgical implantation [33]. Crabbé et al. repopulated the decellularized mouse lungs with two relevant mouse cell types: bone-marrow-derived mesenchymal stromal (stem) cells (MSCs) and alveolar type II cells (C10) in static and a dynamic low fluid shear suspension bioreactor. They reported that dynamic suspension culture is promising for enhancing repopulation of decellularized lungs [48].

Bioengineered Kidney

Song et al. generated and characterized acellular rat kidney scaffolds, repopulated it with epithelial and endothelial cells, then perfused these cell-seeded ECM in a whole organ bioreactor. Finally, following transplantation to rat, the kidney function was evaluated by the ability to produce and excrete urine [47].

Bioengineered Heart

Robertson et al. recellularized whole decellularized hearts with functional endothelial cells, which reduces the thrombogenicity of the scaffold. They showed re-endothelialization improves contractile function of constructs that have been recellularized. Their study was the first step toward generating an engineered functional arterial and venous vasculature that can be used to create transplantable, viable tissues and organs [77]. Lu et al., in a novel strategy, used human iPSCs for recellularization of ECM scaffold and to generate embryoid bodies (EBs) encompassing iPSCs derived multipotential cardiovascular progenitors (MCPs). They added exogenous growth factors such as Bone morphogenetic protein 4 (BMP4), basic fibroblast growth factor (bFGF), Activin-A, Vascular endothelial growth factor A (VEGF), and dickkopf homologue 1 (DKK1) to the culture medium and then seeded the MCPs on ECM scaffold. Results showed that iPSCs-derived MCPs efficiently differentiated into the cardiomyocytes (CMs), smooth muscle cells (SMCs), and endothelial cells (ECs) and exhibited the structural and functional aspects of myocardium and vessel [78].

Bioengineered Pancreas

De Carlo et al. provided acellular pancreatic matrix and demonstrated that islets adhered to the pancreatic matrix showed a constant glucose-induced insulin release during long-term in vitro incubation. Notably, islets cultured without a matrix or on the liver matrix showed a progressive reduction. Totally, they introduced decellularized pancreatic matrix as a suitable scaffold for rat islet cultures, which restored, at least in part, insulin secretion [79]. Conrad et al. recellularized the decellularized pancreatic ECM obtained from rat with human donor islet cells and MSCs. The regenerated pancreas showed preserved subcellular anatomy and insulin response to glucose [80]. In another study, Katsuki et al. performed repopulation of the acellular scaffold with isolated primary porcine islets. Finally, they could observe the cellular integrity of islets and insulin release [81]. Goh et al. generated a whole pancreas ECM using perfusion decellularization technique. They reseeded it with acinar and beta cell lines and inoculated for a 5-day period. Their results showed strong upregulation of insulin gene expression [82].

Altogether, these findings support the biological utility of whole organ ECM as the biomaterial for supporting and enhancing tissue-specific cell functionality and represent a step toward bioengineered organ using regenerative medicine approaches. However, despite the aforementioned advantages, there still exist many complication and challenges that we describe here.

FUTURE PERSPECTIVE AND CHALLENGES OF ORGAN-REGENERATION

Regarding the utilizing of cell-free ECM as a scaffold for organ regeneration and transplantation, there are still many unanswered questions that have caused complications on one side and have caused limitations for clinical use, on the other side.

In several studies, it has been reported that upon transplantation of bioengineered organ, infection, immune response, tissue necrosis, cell proliferation inhibition, and even cytotoxic effect have occurred. The other issue is ECM-cell interaction. It is one of the most important questions that require to be unraveled. Can ECM-cell interaction affect the gene expression patterns? Or how does ECM support the cellular proliferation and differentiation to various cell types?

Nevertheless, we need bioengineered organ models to investigate this complication and address the issues. Of course, it is worth noting that some work has also been done in this regard. For instance, Ross et al. investigated the proliferation and differentiation of the pluripotent murine embryonic SCs by repopulating them on decellularized intact rat kidneys without the addition of any prodifferentiation agents into the culture media. Their results showed that just seeding and adhesion of SCs to the scaffold led to their proliferation and differentiation. Therefore, they introduced a model for studying the signaling mechanisms for growth and differentiation between precursor cells and their matrix surroundings [29].

In addition to the aforementioned questions, there exist issues that are clinically important. Assuming that the transplantation of bioengineered organ is successful and it is long-term viable, is the engineered organ being prone to tumor and fibrosis?

On the other hand, 3D bioprinting technology has recently grown and received an increasing attention. Remarkably, this novel approach can compensate some of the obstacles related to decellularization/recellularization strategy; however, it encounters with several challenges, including bioprinters with high-resolution distribution of cells, controlled deposition of bioinks, suitable bioink formulation, maturation of cells, providing sufficient nutrient and oxygen supply to the engineered living constructs and effective vascularization and innervation within engineered complex structures (reviewed in Refs. [83, 84]).

In parallel with the advances made in the field of tissue engineering, especially robotic-assisted bioprinting technologies, the strong approach of genome editing has emerged. The combination of these two novel approaches to do genetic changes in the SCs, and construct the new tissues through these genetically modified cells anticipates a revolution in future of medical sciences.

On the whole, given these limitations and obstacles in the clinical application of regenerative technologies, the use of bioengineered organs as a definitive method for the treatment of diseases cannot yet be identified, at the present moment. It is, however, at the very beginning of a way that requires a lot of basic and medical studies and evaluations.

In parallel with the advances made in the field of tissue engineering, especially robotic-assisted bioprinting technologies, the strong approach of genome editing has emerged. The combination of these two novel approaches to do genetic changes in the SCs and construct the new tissues through these genetically modified cells anticipates a revolution in future of medical sciences.

REFERENCES

[1] Forostyak S, Jendelova P, Sykova E. The role of mesenchymal stromal cells in spinal cord injury, regenerative medicine and possible clinical applications. Biochimie 2013;95(12):2257–70.
[2] Bartunek J, et al. Cardiopoietic stem cell therapy in heart failure: the C-CURE (cardiopoietic stem cell therapy in heart failure) multicenter randomized trial with lineage-specified biologics. J Am Coll Cardiol 2013;61(23):2329–38.
[3] Lin F, et al. Hematopoietic stem cells contribute to the regeneration of renal tubules after renal ischemia-reperfusion injury in mice. J Am Soc Nephrol 2003;14(5):1188–99.
[4] Santin M. Strategies in regenerative medicine. New York: Springer; 2009.
[5] Sampogna G, Guraya SY, Forgione A. Regenerative medicine: historical roots and potential strategies in modern medicine. J Microsc Ultrastruct 2015;3(3):101–7.
[6] Haseltine WA. The emergence of regenerative medicine: a new field and a new society. J Regener Med 2001;2(4):17–23.
[7] Mamidi MK, et al. Allogeneic and autologous mode of stem cell transplantation in regenerative medicine: which way to go? Med Hypotheses 2014;83(6):787–91.
[8] Koria P. Delivery of growth factors for tissue regeneration and wound healing. BioDrugs 2012;26(3):163–75.
[9] Sethi N. Applications of recombinant protein therapeutic agents in periodontics contributors. Int J Med Med Sci 2013;5(9):380–90.
[10] Ruiz MM, Regueiro JR. New tools in regenerative medicine: gene therapy. Stem Cell Transplant 2012;741:254.
[11] Ayala Garcia MA, et al. The major histocompatibility complex in transplantation. J Transplant 2012;2012:842141.
[12] Beyar R. Challenges in organ transplantation. Rambam Maimonides Med J 2011;2(2):e0049.
[13] Jungebluth P, et al. Tracheobronchial transplantation with a stem-cell-seeded bioartificial nanocomposite: a proof-of-concept study. Lancet 2011;378(9808):1997–2004.
[14] Song JJ, Ott HC. Organ engineering based on decellularized matrix scaffolds. Trends Mol Med 2011;17(8):424–32.
[15] Tapias LF, Ott HC. Decellularized scaffolds as a platform for bioengineered organs. Curr Opin Organ Transplant 2014;19(2):145–52.
[16] Vats A, et al. Scaffolds and biomaterials for tissue engineering: a review of clinical applications. Clin Otolaryngol Allied Sci 2003;28(3):165–72.
[17] Velema J, Kaplan D. Biopolymer-based biomaterials as scaffolds for tissue engineering. Adv Biochem Eng Biotechnol 2006;102:187–238.
[18] O'Brien FJ. Biomaterials & scaffolds for tissue engineering. Mater Today 2011;14(3):88–95.

[19] Yi S, et al. Extracellular matrix scaffolds for tissue engineering and regenerative medicine. Curr Stem Cell Res Ther 2017;12(3):233–46.
[20] Bonnans C, Chou J, Werb Z. Remodelling the extracellular matrix in development and disease. Nat Rev Mol Cell Biol 2014;15(12):786–801.
[21] Humphrey JD, Dufresne ER, Schwartz MA. Mechanotransduction and extracellular matrix homeostasis. Nat Rev Mol Cell Biol 2014;15(12):802–12.
[22] Minton K. Extracellular matrix: preconditioning the ECM for fibrosis. Nat Rev Mol Cell Biol 2014;15(12):766–7.
[23] Mouw JK, Ou G, Weaver VM. Extracellular matrix assembly: a multiscale deconstruction. Nat Rev Mol Cell Biol 2014;15(12):771–85.
[24] Faraj KA, van Kuppevelt TH, Daamen WF. Construction of collagen scaffolds that mimic the three-dimensional architecture of specific tissues. Tissue Eng 2007;13(10):2387–94.
[25] Garcia Y, et al. In vitro characterization of a collagen scaffold enzymatically cross-linked with a tailored elastin-like polymer. Tissue Eng Part A 2009;15(4):887–99.
[26] Kirk JF, et al. Mechanical and biocompatible characterization of a cross-linked collagen-hyaluronic acid wound dressing. Biomatter 2013;3(4):e25633.
[27] Gilbert TW. Strategies for tissue and organ decellularization. J Cell Biochem 2012;113(7):2217–22.
[28] Guyette JP, et al. Perfusion decellularization of whole organs. Nat Protoc 2014;9(6):1451–68.
[29] Ross EA, et al. Embryonic stem cells proliferate and differentiate when seeded into kidney scaffolds. J Am Soc Nephrol 2009;20(11):2338–47.
[30] Liu CX, et al. Preparation of whole-kidney acellular matrix in rats by perfusion. Nan Fang Yi Ke Da Xue Xue Bao 2009;29(5):979–82.
[31] Bolland F, et al. Development and characterisation of a full-thickness acellular porcine bladder matrix for tissue engineering. Biomaterials 2007;28(6):1061–70.
[32] Aubin H, et al. Decellularized whole heart for bioartificial heart. Methods Mol Biol 2013;1036:163–78.
[33] Daly AB, et al. Initial binding and recellularization of decellularized mouse lung scaffolds with bone marrow-derived mesenchymal stromal cells. Tissue Eng Part A 2012;18(1–2):1–16.
[34] Gillies AR, et al. Method for decellularizing skeletal muscle without detergents or proteolytic enzymes. Tissue Eng Part C Methods 2011;17(4):383–9.
[35] Reing JE, et al. The effects of processing methods upon mechanical and biologic properties of porcine dermal extracellular matrix scaffolds. Biomaterials 2010;31(33):8626–33.
[36] Sierad LN, et al. Functional heart valve scaffolds obtained by complete decellularization of porcine aortic roots in a novel differential pressure gradient perfusion system. Tissue Eng Part C Methods 2015;21(12):1284–96.
[37] Guler S, et al. Supercritical carbon dioxide-assisted decellularization of aorta and cornea. Tissue Eng Part C Methods 2017;23(9):540–7.
[38] Huang YH, et al. Preparation of acellular scaffold for corneal tissue engineering by supercritical carbon dioxide extraction technology. Acta Biomater 2017;58:238–43.
[39] Gupta SK, Mishra NC. Erratum to: Decellularization Methods for Scaffold Fabrication. Methods Mol Biol 2018.
[40] Pahk KJ, et al. A novel approach to ultrasound-mediated tissue decellularization and intrahepatic cell delivery in rats. Ultrasound Med Biol 2016;42(8):1958–67.
[41] Xu K, et al. Efficient decellularization for tissue engineering of the tendon-bone interface with preservation of biomechanics. PLoS One 2017;12(2):e0171577.
[42] Kudryk VL, et al. Toxic effect of ethylene-oxide-sterilized freeze-dried bone allograft on human gingival fibroblasts. J Biomed Mater Res 1992;26(11):1477–88.

[43] Gratzer PF, Harrison RD, Woods T. Matrix alteration and not residual sodium dodecyl sulfate cytotoxicity affects the cellular repopulation of a decellularized matrix. Tissue Eng 2006;12(10):2975–83.
[44] Morris AH, Chang J, Kyriakides TR. Inadequate processing of decellularized dermal matrix reduces cell viability in vitro and increases apoptosis and acute inflammation in vivo. Biores Open Access 2016;5(1):177–87.
[45] Freytes DO, et al. Biaxial strength of multilaminated extracellular matrix scaffolds. Biomaterials 2004;25(12):2353–61.
[46] Chi Ting Au-Yeung G, et al. Restoring the biophysical properties of decellularized patches through recellularization. Biomater Sci 2017;5(6):1183–94.
[47] Song JJ, et al. Regeneration and experimental orthotopic transplantation of a bioengineered kidney. Nat Med 2013;19(5):646–51.
[48] Crabbé A, et al. Recellularization of decellularized lung scaffolds is enhanced by dynamic suspension culture. PLoS One 2015;10(5):e0126846.
[49] Garreta E, et al. Tissue engineering by decellularization and 3D bioprinting. Mater Today 2017;20(4):166–78.
[50] Ong CS, et al. Biomaterial-free three-dimensional bioprinting of cardiac tissue using human induced pluripotent stem cell derived cardiomyocytes. Sci Rep 2017;7(1):4566.
[51] Lee V, et al. Design and fabrication of human skin by three-dimensional bioprinting. Tissue Eng Part C Methods 2014;20(6):473–84.
[52] Cui X, et al. Direct human cartilage repair using three-dimensional bioprinting technology. Tissue Eng Part A 2012;18(11–12):1304–12.
[53] Xue SH, et al. Three dimensional bioprinting technology of human dental pulp cells mixtures. Beijing Da Xue Xue Bao 2013;45(1):105–8.
[54] Shim JH, et al. Three-dimensional bioprinting of multilayered constructs containing human mesenchymal stromal cells for osteochondral tissue regeneration in the rabbit knee joint. Biofabrication 2016;8(1):014102.
[55] Bajada S, et al. Updates on stem cells and their applications in regenerative medicine. J Tissue Eng Regen Med 2008;2(4):169–83.
[56] Bongso A, Lee EH. Stem cells: their definition, classification and sources. Stem cells: from bench to bedside. vol. 1. Singapore: World Scientific Publishing; 2005.
[57] Yamanaka S. Induced pluripotent stem cells: past, present, and future. Cell Stem Cell 2012;10(6):678–84.
[58] Zhang G, et al. Induced pluripotent stem cell consensus genes: implication for the risk of tumorigenesis and cancers in induced pluripotent stem cell therapy. Stem Cells Dev 2011;21(6):955–64.
[59] Witkowska-Zimny M, Wrobel E. Perinatal sources of mesenchymal stem cells: Wharton's jelly, amnion and chorion. Cell Mol Biol Lett 2011;16(3):493.
[60] Pappa KI, Anagnou NP. Novel sources of fetal stem cells: where do they fit on the developmental continuum? Regen Med 2009;4(3):423–33.
[61] Bajek A, et al. Human amniotic-fluid-derived stem cells: a unique source for regenerative medicine. Expert Opin Biol Ther 2014;14(6):831–9.
[62] De Coppi P, et al. Isolation of amniotic stem cell lines with potential for therapy. Nat Biotechnol 2007;25(1):100–6.
[63] Finegan JAK. Amniotic fluid and midtrimester amniocentesis: a review. Br J Obstet Gynaecol 1984;91(8):745–50.
[64] Phermthai T, et al. A novel method to derive amniotic fluid stem cells for therapeutic purposes. BMC Cell Biol 2010;11(1):79.

[65] Tsai MS, et al. Isolation of human multipotent mesenchymal stem cells from second-trimester amniotic fluid using a novel two-stage culture protocol. Hum Reprod 2004;19(6):1450–6.

[66] Roselli EA, et al. Fetal mesenchymal stromal cells from cryopreserved human chorionic villi: cytogenetic and molecular analysis of genome stability in long-term cultures. Cytotherapy 2013;15(11):1340–51.

[67] Fauza D. Amniotic fluid and placental stem cells. Best Pract Res Clin Obstet Gynaecol 2004;18(6):877–91.

[68] Arakawa R, et al. Human first-trimester chorionic villi have a myogenic potential. Cell Tissue Res 2012;348(1):189–97.

[69] Jansson T. Placenta plays a critical role in maternal–fetal resource allocation. Proc Natl Acad Sci USA 2016;113(40):11066–8.

[70] Fernandes RA, et al. Placental tissues as sources of stem cells. Open J Anim Sci 2012;2(03):166.

[71] Katsiani E, et al. Chorionic villi derived mesenchymal like stem cells and expression of embryonic stem cells markers during long-term culturing. Cell Tissue Bank 2016;17(3):517–29.

[72] Pipino C, et al. Placenta as a reservoir of stem cells: an underutilized resource? Br Med Bull 2013;105(1):43–68.

[73] Alfirevic Z, Sundberg K, Brigham S. Amniocentesis and chorionic villus sampling for prenatal diagnosis. Cochrane Database Syst Rev 2003(3):CD003252.

[74] Nagamura-Inoue T, He H. Umbilical cord-derived mesenchymal stem cells: their advantages and potential clinical utility. World J Stem Cells 2014;6(2):195.

[75] Wang HS, et al. Mesenchymal stem cells in the Wharton's jelly of the human umbilical cord. Stem Cells 2004;22(7):1330–7.

[76] Malgieri A, et al. Bone marrow and umbilical cord blood human mesenchymal stem cells: state of the art. Int J Clin Exp Med 2010;3(4):248.

[77] Robertson MJ, et al. Optimizing recellularization of whole decellularized heart extracellular matrix. PLoS One 2014;9(2):e90406.

[78] Lu TY, et al. Repopulation of decellularized mouse heart with human induced pluripotent stem cell-derived cardiovascular progenitor cells. Nat Commun 2013;4:2307.

[79] De Carlo E, et al. Pancreatic acellular matrix supports islet survival and function in a synthetic tubular device: in vitro and in vivo studies. Int J Mol Med 2010;25(2):195–202.

[80] Conrad C, et al. Bio-engineered endocrine pancreas based on decellularized pancreatic matrix and mesenchymal stem cell/islet cell coculture. J Am Coll Surg 2010;211(3):S62.

[81] Katsuki Y, Yagi H, Okitsu T, Kitago M, Tajima K, Kadota Y, Hibi T, Abe Y, Shinoda M, Itano O, et al. Endocrine pancreas engineered using porcine islets and partial pancreatic scaffolds. Pancreatology: Int J Pancreatol 2016;16(5):922–30.

[82] Goh SK, et al. Perfusion-decellularized pancreas as a natural 3D scaffold for pancreatic tissue and whole organ engineering. Biomaterials 2013;34(28):6760–72.

[83] Mir TA, Nakamura M. Three-dimensional bioprinting: toward the era of manufacturing human organs as spare parts for healthcare and medicine. Tissue Eng Part B Rev 2017;23(3):245–56.

[84] Zhang HB, et al. Three-dimensional bioprinting is not only about cell-laden structures. Chin J Traumatol 2016;19(4):187–92.

Chapter 6

Introduction to Transplantology

Emilia Wiecheć*, Jolanta Hybiak†, Claudine Kieda‡
*Department of Clinical and Experimental Medicine (IKE), Division of Cell Biology, Linköping University, Linköping, Sweden
†Department of Pathomorphology, Pomeranian Medical University in Szczecin, Szczecin, Poland
‡Department of Oncology with Laboratory of Molecular Oncology, Military Institute of Medicine, Warsaw, Poland

INTRODUCTION

The chapter provides a historic perspective of transplantology and highlights the key events important for the development of the field. It approaches the topic from both immunomedical perspective, as well as it highlights how immunology benefited from transplantology and how the newest developments in immunology allowed for the development of better transplantation protocols. The terms, autotransplant, allotransplant and xenotransplant, chimera, and induced tolerance are explained and several examples for each type of transplantation procedure are provided. The historic milestones in the transplantation of kidneys, skin, pancreatic islets and liver that support the progress in transplantology are described. The final parts of the chapter are dedicated to vascularized composite allografts that cover, that is, facial transplants, hand transplants, etc. as well as the basics of regenerative medicine.

XENOGRAFTS

According to the International Xenotransplantation Association, "xenotransplantation refers to procedures in which live cells, tissues or organs derived an animal are transplanted, implanted, or infused into human" [1]. The first known animal to human organ transplantations by vascular anastomosis was carried out at the beginning of the 20th century. The first attempts were made with kidney, using as a donor pig, goat [2], lamb [3], and nonhuman primate [4]. None of them were successful, majority of the transplanted kidney did not function at all and human recipients died within few hours. The longest survival period after renal xenotransplantation was 9 days [3]. In 1963 Reemtsma and his group due to the lack of living or deceased donor organs decided to transplant kidney from chimpanzees to six patients with terminal renal disease [5]. Recipients underwent an immunosuppressive therapy

Stem Cells and Biomaterials for Regenerative Medicine. https://doi.org/10.1016/B978-0-12-812258-7.00006-X

prior and after operation. Most of the patients died within 2 months except one who gained a normal renal function and survived 9 months dying eventually of an electrolyte disturbance. Transplant surgeons encouraged by Reemtsma experience began working with clinical xenografting. In 1964 Hume did a chimpanzee-to-man renal transplantation, but the patient died the next day due to the excessive diuresis [6]. One year later, in 1964 Thomas E. Starzl carried out baboon kidney transplantation in six patients with terminal uremia. He reported that grafts survived from 19 to 60 days and were lost due to continuous complement activation [7]. The development of a cadaver organ procurement and availability of hemodialysis facility led to discontinuation of the clinical kidney xenotransplantation [8]. Between 1966 and 1974, Starzl carried out one ex vivo liver perfusion [9] and orthotopic liver transplants with chimpanzees as organs donors [10, 11], all with a fatal outcome. In 1992 and 1993, he carried out transplantation in two patients, of which one survived 70 days [12]. The first case of chimpanzee's heart transplantation in man was reported by Hardy in 1964 [13], however unsuccessful one. Later, between 1968 and 1992, few transplantation attempts of an animal heart to human took place [14–17]. Most patients did not even survive 1 day, the longest survival period was 20 days [16]. The first pig islet transplant was reported in 1994 by Groth [18]. The C-peptide was detected in patient's urine proving that porcine pancreatic endocrine tissue can survive in human body, but no clinical benefits were achieved. Another example of porcine tissue transplantation is the clinical trial of intraperitoneal alginate-poly-L-ornithine-alginate (APA) encapsulated (neonatal) porcine islet xenotransplantation in nonimmunosuppressed diabetic patients. This project was carried out from August 2011 to July 2012 in New Zealand and showed a significant reduction of the frequency of unaware hypoglycemic events after transplantation [19].

The xenotransplantation of organs human requires overcoming not only the immunological barrier, which is much more severe and difficult to control than the immune response to allografts, [20] as well as physiologic limitations and animal diseases transmission (xenozoonosis). The disease transmission from animals to humans is according to WHO the main risk in xenotransplantation [21]. It is predicted that the number of available human organs will never meet the demand [22, 23]. Nowadays, special attention is paid to pigs and introducing genetic changes into these animals [24–26]. Pigs possess important characteristics that make them suitable organ donors for human. Pigs are appropriate in size and are readily bred, which makes them available in large numbers and simplifies genetic engineering work. Moreover, pigs can be raised under strict conditions and free of known pathogens [20, 27]. In recent years, the genetic engineering has been used to generate pigs expressing multiple human genes with knocked out pig's genes involved in hyperacute rejection [28, 29] or even a human-pig chimera [30]. Genetic modifications combined with new pharmacological immunosuppressants resulted in extended survival of nonhuman primates who received receives pig-derived transplantation [24, 31–33]. However, scientists involved in xenotransplantation development claim that "industry appears reluctant to support research that will take more than 2–3 years to come to clinical trial, and the funding agencies appear to have been "distracted" by the current appeal of stem cell technology and regenerative medicine" [34].

The shortage of human donors pushed scientists to search for another source of organs. Xenotransplantation was an attractive alternative; unfortunately, shortly the immunological and physiological barriers between human and other species had occurred making this procedure complicated. It seems that without genetic engineering, xenotransplantation has no chance to be successfully applied in the human transplantology. Furthermore, xenotransplantation may raise many ethical concerns.

AUTOGRAFTS

The term autograft refers to the transplant that comprises individual's own organ, tissue, or cells, transferred from one part of the body to another. The first report of an experimental autologous skin transplantation was published in 1804 [35]. Giuseppe Baronio resected a piece of skin of a ram and grafted it on the other side of the body in the same animal. He carried out a series of experiments on the same ram with different periods of time between excision and grafting. He noticed that the shorter time gap between skin harvesting and grafting, the better outcome. Later, in the XIX century, human skin autotransplantations were performed also as treatment for wounds and burned skin [36, 37]. Technique of skin grafting developed over the next years, giving the principal rules for current plastic surgery [38, 39]. Nowadays autologous skin grafts are commonly considered as a gold standard for treatment of defects after surgical resection, burn, chronic ulcers, scar contracture release, hair restoration, or vitiligo [40–42]. In 1821, Philipp Franz von Walther reconstructed a part of skull, which was the first autograft bone operation. Today, orthopedic reconstructive surgery uses bone autografting due to its biologic and mechanical properties. Bones comprise a natural mineral and reservoir of growth factors as well as possess three crucial features such as osteogenic, osteoinductive, and osteoconductive, hence facilitating shorter wound healing and bone formation [43, 44]. In 1905 Carrel and Guthrie carried out dog's kidney transplantation. They noticed that kidney taken from and grafted into the same animal survived and worked [45]. Other examples of autografts in surgery come from anterior cruciate ligament repair [46], fat autografts in soft-tissue defects in face, trunk, extremities correction including breast reconstruction [47], autologous blood transfusion of the previously donated blood carried out during or after operation [48]. The advantage of the autografts over other types of grafts is the lack of antigen–antibody reaction and high success rate in treatment, with no risk of disease transmission [49]. However, donor's morbidity, the limited volume available especially in pediatric and elderly patients, requirement of additional surgical site and additional pain [50], higher blood loss and need for transfusion [44], must be taken into consideration.

ALLOGRAFTS

In 1871, George Pollock described that skin graft taken from and transplanted to the same patient were successful, while grafts derived from other donors failed. At the same time, he distinguished autografts from allografts [51]. Unfortunately, his report was ignored and the reason of graft tolerance remained unexplained for other decades [52]. In 1912, Schöne not only confirmed that allografts always fade but also went a step further. He noticed that subsequent grafts from the same donor failed even faster than the first one [53]. He is called the first transplantation immunologist. Soon, other scientists proved the role of lymphocytes in grafts destruction. By means of irradiation, benzol, or splenectomy, they reduced lymphocyte infiltration usually observed in failed allografts, and at the same time extended graft survival period [54, 55]. In 1933 Voronoy carried out the first human-to-human kidney allotransplantation, however, not successful one [56]. Unexpected results of Medawar and Billingham's experiments shed light on allotransplantation [57]. Their skin allografting experiment performed on identical (monozygotic) and fraternal (dizygotic) cattle twins showed that most cows accepted their twin's graft even if they were identical or fraternal twins. Contrary to twin grafts, skin grafts interchanged between animals of the other degrees of relationship were rapidly destroyed. With the support of previous observations by Hunter, Lillie [58], and Owen [59], they concluded that the twin cattle exchange in the utero not only hormones and red blood cells but also leukocytes, which are responsible for graft tolerance. Medawar, Billingham, and Brent induced chimerism and allograft acceptance in mice by inoculating intrauterine fetuses with donor strain spleen cells. In adulthood, the mice accepted skin allografts from the donor strain [60], giving evidence of allograft likelihood without rejection. In 1954, the first successful human kidney transplantation was performed between identical twins [61]. Although induction of chimerism in human recipients by neonatal treatment was clearly impossible, surgeons encouraged by successful human-to-human transplantation were pushed to look for a new solution. One year later it was shown that inoculating bone marrow cells into adult mice with impaired immunological response after irradiation could induce chimerism and those mice accepted skin grafts from bone marrow donors [62]. In 1958, Murray performed a similar procedure on 12 patients, who were treated with total body irradiation and two of them got a donor bone marrow transplanted [63]. Unfortunately, only one patient survived and afterwards received kidney of his fraternal twin brother, which kept its functions for 20 years. It was the first case of successful transplantation between nonidentical twins. Another two successful kidney transplantations from nontwin donors and after recipient's total irradiation were carried out with immunosuppressant aid (steroids and 6-mercaptopurine). These cases provide a firm evidence that close familial relationship is not essential for a long survival of the transplant [64]. Meanwhile, investigations on drug-based immunosuppression were being conducted. The first reports showed that this kind of drugs reduced the antibody response to bovine albumen [65], prolonged the survival time of skin allograft in rabbit [66], inhibited dog kidney allograft rejection [67]. Despite hopes associated with immunosuppressive drugs, outcomes in human organ transplantation were rather disappointing, no significant survival was achieved, and patient usually died within 6 months [68]. In 1963, the huge step forward was made after Starzl introduced his approach to immunosuppression with satisfactory results [69]. Furthermore, advanced analysis method enabled to detect donor lymphocytes in skin biopsies, lymph nodes, and other tissues. Later Dr. Thomas Starzl, a pioneer and a great authority in transplantology, concluded that there were two decisive turning points in the evolution of clinical transplantology. The first one was the induction of chimerism-associated neonatal tolerance and the second was the demonstration that organ allografts could self-induce tolerance with the aid of immunosuppression [70]. With time the definition of death has evolved from the stop of heart beating to the loss of brain function (brain dead). The first removal of kidneys from the brain dead and heart-beating cadaver was performed in 1963 [71]. In 1968, the irreversible loss of brain function was accepted as a death in a recommendation released by the "Harvard ad hoc Committee on Brain Death" [72]. It took some time before the transplant form brain dead donor with still beating heart gained wide acceptance. This approach increased availability and quality of organs from decayed donors due to reduction of ischemic damage in comparison to nonheart-beating cadaver donors [73]. Advancement in the field of organ preservation with lower temperature and perfusing solution enabled procurement of multiple organs and extended their suitability for transplantation even till 3 days [74–76]. Discovery of human leukocyte antigen (HLA) [77, 78] led to the development of a microcytotoxicity assay by Terasaki [79], which is about mixing serum of the recipient with donor lymphocytes in order to select appropriate donors as well as predicting the possibility of hyperacute rejection.

With the accumulation of knowledge within transplantation immunology field and a growing experience with kidney transplantation, the next important target for transplantologists was liver. Mastering liver transplantology involved series of attempts performed on animals and humans. Between 1963 and 1964, seven liver transplantations (Boston, Denver, and Paris) from deceased donors were carried out. All patients died within 23 days due to postsurgery bleeding, pulmonary emboli, or sepsis. Neither primary dysfunction nor rejection of the grafts was a lethal factor in any of the failure. Such a bad outcome led to a voluntary worldwide moratorium to stop liver transplantations. Liver transplantation seemed at that time too difficult and extremely expensive [80]. Long-term liver replacement therapy was beyond reach of medicine due to the high diversity of liver function. Moratorium time brought a bloom of hepatotrophic physiology into multiple research areas of metabolism and regenerative medicine to get better understanding of the structure, role, and function of liver. Dual blood supplies, venous drainage, and biliary system make a surgical connection of the transplanted liver a huge challenge. Animal model helped to develop surgical methods and overcome obstacles. Moreover, clinical introduction of antilymphoid globulin [81] and growing awareness of importance of HLA matching in organ transplantation [82] led to improvement in human liver transplantation. Between 1967 and 1968, Starzl and his team carried out seven orthotopic liver transplantations, during which all patients passed through operative and postoperative period, previously related with high mortality. They improved liver transplantation by histocompatibility profiling, the application of an effective organ preservation system, and the use of an immunosuppressive regiment. Their work was appreciated as absolutely outstanding and magnificent achievement [83]. By 1983, over 540 liver transplantations had been carried out in four medical centers in the United States and Western Europe. Obtained results demonstrated that liver transplantation offers an alternative therapeutic approach for patients with severe liver disease. Finally, in 1983 in Bethesda, the NIH Consensus Panel consisting of herpetologist, surgeon, internist, pediatricians, immunologists, biostatisticians, ethicists, and public representatives declared liver transplantation as a therapeutic alternative, leading to a spread of the practice worldwide [84].

The main problems connected with allotransplantation such as immunosuppression, tissue matching, and organ preservation, after years of attempts, were worked out mostly with the kidney and liver, which allowed to apply with the minor modification transplantation procedures to other organs. Transplantology is considered a wide field and transplantation of a particular organ comprising historical overview and surgical procedures is a separate branch, for example, heart [85], lung [86], pancreatic islet [87], and intestine [88].

Nowadays, such complex transplantation as hand, forearm, arm and partial- or full-face transplantation to patient after severe injuries or diseases are performed. Such grafts are called vascularized composite allografts (VCAs). The unique value of this procedure is restoring function, esthetic, social, and psychological outcome [89]. The patient will survive without the transplant, so it is not lifesaving but a life giving/improving procedure [90]. The indication for allotransplantation must be still restrictive, due to the need for immunosuppressive protection with all associated side effects of this treatment [91]. The era of VCAs began in 1998 with the first successful hand allotransplantation performed in 1998 from a brain-dead man aged 41 years to a man aged 48 years who had had traumatic amputation of the distal third of his right forearm. In the report, it was concluded that hand transplantation is technically feasible, current immunosuppression protocols prevent from acute rejection, and since good long-term outcome is achievable, it might be similar to if not better than autoreconstruction [92]. The most complex microvascular reconstruction procedure available nowadays is facial allotransplantation. Maria Siemionow's team in United States had been reporting animal models of face transplantation since the beginning of 2000s and explored the procedure in a nonclinical setting before their first face allotransplantation in 2008 [93]. However, the first ever partial facial transplantation was carried out in 2005 in France [94], and the first full-face transplant was performed in Barcelona in 2010 [95]. Till 2016, at least

35 face allotransplantations were performed worldwide. There were at least five deaths reported due to surgical complication or tumor development related to immunosuppression [96]. Hence, such procedures still involve a very high risk and cannot be named a routine surgery [97].

Allotransplantation currently remains the treatment of choice for patient with for example severely advanced heart failure, end-stage liver failures when no other treatment option, such as medical therapy or electrical and mechanical devices can compete with the expected good long-term results. However, rejection and the consequences of immunosuppressive therapy still constitute a problem, which limits the success of transplantation. Moreover, number of patients with severely advanced organ failure increases, while at the same time number of available organs is limited. Allotransplantation remains an option only for a limited number of patients.

GRAFTS INVOLVING ARTIFICIAL BIOMATERIALS

American National Institute of Health defines *biomaterials* as "any substance or combination of substances, other than drugs, synthetic or natural in origin, which can be used for any period of time, and augments or replaces partially or totally any tissue, organ or function of the body, in order to maintain or improve the quality of life of the individual." The artificial material for tissue repair is used for example as a substitute of skin [98], acellular vascular tissue [99], bone [43] or in urinary tract [100], and cornea [101] reconstruction. Although artificial biomaterials do not contain foreign proteins and do not elicit an immune response from the host, no man-made material is capable of self-repair or adaptation and is limited in function. Because of these limitations, tissue engineering and nanotechnology aim to replace lost tissue or organs with polymeric scaffolds that mimic three-dimensional natural extracellular matrix and contain specialized population of living cells to regenerate tissue and restore normal organ function. Ideally, degradation products of such polymers are nontoxic, capable of entering normal metabolic pathways, and due to lack of biologically functional domains, are of low immunogenicity. A tensile strength and degradation rate are properties that can be easily modified and adjusted to the demands of particular application by changing parameters and ingredients used in polymerization process. Such tissue-engineered construct of a scaffold with organ-specific cells may be implanted and promote growth and development of new tissues. Tissue-engineered constructs have been used successfully in bone [102], skin [103], and bladder [104].

REFERENCES

[1] Schmidt M, et al. Spheroid-based 3-dimensional culture models: gene expression and functionality in head and neck cancer. Oncol Rep 2016;35(4):2431–40.

[2] Jaboulay M. Greffe du reins au pli du coude par soudures arterielles et veineuses. (Kidney grafts in the antecubital fossa by arterial and venous anastomosis.). Lyon Med 1906;107:575–7.

[3] Neuhof H, Hirshfeld S. The transplantation of tissues. New York: Appleton; 1923.
[4] Unger E. Nierentransplantation. (Kidney transplantation.). Wien Klin Wochenschr 1910;47:573–8.
[5] Reemtsma K, et al. Heterotransplantation of the kidney: two clinical experiences. Science 1964;143(3607):700–2.
[6] Hume DM, et al. Studies of renal homotransplantation in man. Ann N Y Acad Sci 1964; 120(1):578–606.
[7] Starzl TE, et al. Renal heterotransplantation from baboon to man: experience with 6 cases. Transplantation 1964;2:752–76.
[8] Reemtsma K. Xenotransplantation: a historical perspective. ILAR J 1995;37(1):9–12.
[9] Starzl TE, et al. Avenues of future research in homotransplantation of the liver with particular reference to hepatic supportive procedures, antilymphocyte serum, and tissue typing. Am J Surg 1966;112(3):391–400.
[10] Starzl TE. Orthotopic heterotransplantation. In: Experience in hepatic transplantation. Philadelphia: WB Saunders Company; 1969. p. 408–21.
[11] Giles GR, et al. Clinical heterotransplantation of the liver. Transplant Proc 1970;2(4):506–12.
[12] Starzl TE, et al. Baboon-to-human liver-transplantation. Lancet 1993;341(8837):65–71.
[13] Hardy JD, et al. Heart transplantation in man—developmental studies + report of case. JAMA, J Am Med Assoc 1964;188(13):1132–40.
[14] Cooley DA, et al. Human heart transplantation. Am J Cardiol 1968;22(6):804–10.
[15] Barnard CN, Wolpowitz A, Losman JG. Heterotopic cardiac transplantation with a xenograft for assistance of the left heart in cardiogenic shock after cardiopulmonary bypass. S Afr Med J 1977;55:1035–8.
[16] Bailey LL, et al. Baboon-to-human cardiac xenotransplantation in a neonate. JAMA 1985;254(23):3321–9.
[17] Czaplicki J, Blońska B, Religa Z. The lack of hyperacute xenogeneic heart transplant rejection in a human. J Heart Lung Transplant 1992;11:393–7.
[18] Groth CG, et al. Transplantation of porcine fetal pancreas to diabetic patients. Lancet 1994;344(8934):1402–4.
[19] Matsumoto S, et al. Clinical benefit of islet xenotransplantation for the treatment of type 1 diabetes. EBioMedicine 2016;12:255–62.
[20] Cascalho M, Platt JL. The immunological barrier to xenotransplantation. Immunity 2001;14(4):437–46.
[21] Belmokhtar CA, Hillion J, Segal-Bendirdjian E. Staurosporine induces apoptosis through both caspase-dependent and caspase-independent mechanisms. Oncogene 2001;20(26):3354–62.
[22] Röck D, Petersen P, Yoeruek E, Thaler S, Bartz-Schmidt KU, Röck T. Effect of organ scandal on corneal donation rate and organ donors at a German University Hospital. Ann Transplant 2017;22:425–30.
[23] Johnson RJ, et al. Organ donation and transplantation in the UK—the last decade: a report from the UK National Transplant Registry. Transplantation 2014;97:S1–27.
[24] Schuurman H-J. Pig-to-nonhuman primate solid organ xenografting: recent achievements on the road to first-in-man explorations. Xenotransplantation 2016;23(3):175–8.
[25] Nagashima H, Matsunari H. Growing human organs in pigs—a dream or reality? Theriogenology 2016;86(1):422–6.
[26] Cooper DKC, et al. The role of genetically engineered pigs in xenotransplantation research. J Pathol 2016;238(2):288–99.
[27] Denner J. Xenotransplantation—a special case of one health. One Health 2017;3:17–22.
[28] Klymiuk N, et al. Genetic modification of pigs as organ donors for xenotransplantation. Mol Reprod Dev 2010;77(3):209–21.

[29] Niemann H, Petersen B. The production of multi-transgenic pigs: update and perspectives for xenotransplantation. Transgenic Res 2016;25(3):361–74.
[30] Wu J, et al. Interspecies chimerism with mammalian pluripotent stem cells. Cell 2007; 168(3):473–86.
[31] Cooper DKC, et al. Progress in pig-to-non-human primate transplantation models (1998–2013): a comprehensive review of the literature. Xenotransplantation 2014;21(5):397–419.
[32] Denner J. Recent progress in xenotransplantation with emphasis on virological safety. Ann Transplant 2016;21:717–27.
[33] Cooper DKC. Early clinical xenotransplantation experiences—an interview with Thomas E. Starzl, MD, PhD. Xenotransplantation 2017;24(2):e12306.
[34] Cooper DKC. The case for xenotransplantation. Clin Transplant 2015;29(4):288–93.
[35] Baronio G. Degli innesti animali. Milan: Stamperia e Fonderia del Genio; 1804.
[36] Freshwater MF, Krizek TJ. George David Pollock and the development of skin grafting. Ann Plast Surg 1978;1(1):96–104.
[37] Lawson G. On the transplantation of portions of skin for the closure of large granulating surfaces. Trans Clin Soc Lond 1871;4:49.
[38] Blair VP, Brown JB. The use and uses of large split skin grafts of intermediate thickness. Plast Reconstr Surg 1968;42(1):65–75.
[39] Reverdin J. Sur la greffe épidermique. Arch Med General Paris 1872;19:276–303.
[40] Shimizu R, Kishi K. Skin graft. Plast Surg Int 2012;2012:563493.
[41] Serra R, et al. Skin grafting for the treatment of chronic leg ulcers—a systematic review in evidence-based medicine. Int Wound J 2017;14(1):149–57.
[42] Valencia IC, Falabella AF, Eaglstein WH. Skin grafting. Dermatol Clin 2000;18(3):521–32.
[43] Campana V, Milano G, Pagano E. Bone substitutes in orthopaedic surgery: from basic science to clinical practice. J Mater Sci Mater Med 2014;25(10):2445–61.
[44] Zhang Y-H, et al. Structural allograft versus autograft for instrumented Atlantoaxial fusions in pediatric patients: radiologic and clinical outcomes in series of 32 patients. World Neurosurg 2017;105:549–56.
[45] Carrel A, Guthrie CC. Functions of a transplanted kidney. Science 1905;22(563):473.
[46] Kiapour AM, Murray MM. Basic science of anterior cruciate ligament injury and repair. Bone Joint Res 2014;3(2):20–31.
[47] Simonacci F, et al. Procedure, applications, and outcomes of autologous fat grafting. Ann Med Surg 2017;20:49–60.
[48] Zhou J. A review of the application of autologous blood transfusion. Braz J Med Biol Res 2016;49(9):e5493.
[49] Miller CP, Chiodo CP. Autologous bone graft in foot and ankle surgery. Foot Ankle Clin 2016;21(4):825–37.
[50] Baumhauer J, et al. Site selection and pain outcome after autologous bone graft harvest. Foot Ankle Int 2014;35(2):104–7.
[51] Pollock G. Cases of skin grafting and skin transplantation. Trans Clin Soc Lond 1871;4: 37–47.
[52] Barker CF, Markmann JF. Historical overview of transplantation. Cold Spring Harb Perspect Med 2013;3(4):a014977.
[53] Schöne G. Die heteroplastische und homooplastische. Berlin: Springer; 1912.
[54] Murphy JB. Factors of resistance to heteroplastic tissue-grafting: studies in tissue specificity. III. J Exp Med 1914;19(5):513–22.
[55] Murphy JB. Heteroplastic tissue grafting effected through roentgen-ray lymphoid destruction. J Am Med Assoc 1914;LXII(19):1459.

[56] Matevossian E, et al. Surgeon Yurii Voronoy (1895–1961)—a pioneer in the history of clinical transplantation: In memoriam at the 75th anniversary of the first human kidney transplantation. Transpl Int 2009;22(12):1132–9.
[57] Anderson D, et al. The use of skin grafting to distinguish between monozygotic and dizygotic twins in cattle. Heredity 1951;5(3):379–97.
[58] Lillie FR. The theory of the free-Martin. Science 1916;43(1113):611–3.
[59] Owen RD. Immunogenetic consequences of vascular anastomoses between bovine twins. Science 1945;102(2651):400–1.
[60] Billingham RE, Brent L, Medawar PB. Pillars article: 'Activity acquired tolerance' of foreign cells *Nature* 172: 603–606. J Immunol 2010;184(1):5–8.
[61] Murray JE, Merrill JP, Harrison JH. Renal homotransplantation in identical twins. Surg Forum 1955;6:432–6.
[62] Main JM, Prehn RT. Successful skin homografts after the administration of high dosage X radiation and homologous bone marrow. J Natl Cancer Inst 1955;15(4):1023–9.
[63] Murray JE, et al. Kidney transplantation in modified recipients. Ann Surg 1962;156(3):337–55.
[64] Kuss R, et al. Homologous human kidney transplantation: experience with six patients. Postgrad Med J 1962;38(443):528–31.
[65] Schwartz R, Dameshek W. Drug-induced immunological tolerance. Nature 1959; 183(4676):1682–3.
[66] Schwartz R, Dameshek W, Donovan J. The effects of 6-mercaptopurine on homograft reactions. J Clin Investig 1960;39(6):952–8.
[67] Calne RY. The rejection of renal homografts. Lancet 1960;275(7121):417–8.
[68] Murray JE, et al. Prolonged survival of human-kidney homografts by immunosuppressive drug therapy. N Engl J Med 1963;268(24):1315–23.
[69] Starzl TE, Marchioro TL, Waddell WR. The reversal of rejection in human renal homografts with subsequent development of homograft tolerance. Surg Gynecol Obstet 1963;117:385–95.
[70] Starzl TE. History of clinical transplantation. World J Surg 2000;24(7):759–82.
[71] Machado C. The first organ transplant from a brain-dead donor, in brain death: a reappraisal. New York, NY: Springer New York; 2007. p. 21–31.
[72] Anon. A definition of irreversible coma: report of the ad hoc committee of the harvard medical school to examine the definition of brain death. JAMA 1968;205(6):337–40.
[73] Kumar L. Brain death and care of the organ donor. J Anaesthesiol Clin Pharmacol 2016;32(2):146–52.
[74] Marchioro TL, et al. Extracorporeal perfusion for obtaining postmortem homografts. Surgery 1963;54:900–11.
[75] Belzer F, Ashby BS, Dunphy JE. 24-Hour and 72-hour preservation of canine kidneys. Lancet 1967;290(7515):536–9.
[76] Belzer FO, et al. The use of UW solution in clinical transplantation a 4-year experience. Ann Surg 1992;215(6):579–85.
[77] Snell GD. Pillars article: methods for the study of histocompatibility genes. *J. Genet.* 1948. 49: 87–108. J Immunol 2014;192(1):5–26.
[78] Dausset J. Iso-leuco-anticorps. Acta Haematol 1958;20(1–4):156–66.
[79] Terasaki PI, McClelland JD. Microdroplet assay of human serum cytotoxins. Nature 1964;204(4962):998–1000.
[80] Starzl TE, Fung JJ. Themes of liver transplantation. Hepatology (Baltimore, Md) 2010;51(6):1869–84.
[81] Starzl TE, et al. The use of heterologous antilymphoid agents in canine renal and liver homotransplantation and in human renal homotransplantation. Surg Gynecol Obstet 1967;124(2):301–8.

[82] Starzl TE, et al. Long-term survival after renal transplantation in humans: (with special reference to histocompatibility matching, thymectomy, homograft glomerulonephritis, heterologous ALG, and recipient malignancy). Ann Surg 1970;172(3):437–72.
[83] Starzl TE, et al. Orthotopic homotransplantation of the human liver. Ann Surg 1968;168(3):392–415.
[84] Anon. Liver transplantation. NIH Consens Statement 1983 Jun 20–23. Available from: https://consensus.nih.gov/1983/1983livertransplantation036html.htm; 1983. (Accessed July 26, 2017).
[85] DiBardino DJ. The history and development of cardiac transplantation. Tex Heart Inst J 1999;26(3):198–205.
[86] Idrees JJ, Pettersson GB. State of the art of combined heart-lung transplantation for advanced cardiac and pulmonary dysfunction. Curr Cardiol Rep 2016;18(4):36.
[87] Ramírez-Domínguez M. In: Ramírez-Domínguez M, editor. Historical background of pancreatic islet isolation, in pancreatic islet isolation: from the mouse to the clinic. Cham: Springer International Publishing; 2016. p. 1–9.
[88] Margreiter R. The history of intestinal transplantation. Transplant Rev 1997;11(1):9–21.
[89] Maciejewski A, et al. The first immediate face transplant in the world. Ann Surg 2016;263(3):e36–9.
[90] Jones NF. Concerns about human hand transplantation in the 21st century. J Hand Surg (Br) 2002;27(5):771–87.
[91] Salminger S, et al. Functional and psychosocial outcomes of hand transplantation compared with prosthetic fitting in below-elbow amputees: a multicenter cohort study. PLoS One 2016;11(9):e0162507.
[92] Dubernard J-M, et al. Human hand allograft: report on first 6 months. Lancet 1999;353(9161):1315–20.
[93] Siemionow M, et al. Near-total human face transplantation for a severely disfigured patient in the USA. Lancet 2009;374(9685):203–9.
[94] Dubernard J-M, et al. Outcomes 18 months after the first human partial face transplantation. N Engl J Med 2007;357(24):2451–60.
[95] Barret JP, et al. Full face transplant: the first case report. Ann Surg 2011;254(2):252–6.
[96] Khalifian S, et al. Facial transplantation: the first 9 years. Lancet 2014;384(9960):2153–63.
[97] Giatsidis G, Sinha I, Pomahac B. Reflections on a decade of face transplantation. Ann Surg 2017;265(4):841–6.
[98] Halim AS, Khoo TL, Mohd Yussof SJ. Biologic and synthetic skin substitutes: an overview. Indian J Plast Surg 2010;43(Suppl):S23–8.
[99] Schmidt CE, Baier JM. Acellular vascular tissues: natural biomaterials for tissue repair and tissue engineering. Biomaterials 2000;21(22):2215–31.
[100] Elsawy MM, de Mel A. Biofabrication and biomaterials for urinary tract reconstruction. Res Rep Urol 2017;9:79–92.
[101] Akpek EK, et al. Artificial corneas versus donor corneas for repeat corneal transplants. Cochrane Database Syst Rev 2014;11:CD009561.
[102] Schimming R, Schmelzeisen R. Tissue-engineered bone for maxillary sinus augmentation. J Oral Maxillofac Surg 2004;62(6):724–9.
[103] Yannas IV, et al. Synthesis and characterization of a model extracellular matrix that induces partial regeneration of adult mammalian skin. Proc Natl Acad Sci U S A 1989;86(3):933–7.
[104] Atala A, et al. Tissue-engineered autologous bladders for patients needing cystoplasty. Lancet 2006;367(9518):1241–6.

Chapter 7

Biomaterials, Definition, Overview

Andrzej Hudecki*, Gerard Kiryczyński†, Marek J. Łos‡,§,¶
*Institute of Non-Ferrous Metals, Gliwice, Poland
†Department of Pathomorphology, Pomeranian Medical University in Szczecin, Szczecin, Poland
‡Biotechnology Centre, Silesian University of Technology, Gliwice, Poland
§LinkoCare Life Sciences AB (Ltd), Linköping, Sweden
¶Center for Molecular Biophysics, UPR4301 CNRS CS80054, Orleans Cedex 2, France

INTRODUCTION

"Biomaterials" journal defines biomaterial as a substance that has been engineered to take form, which, alone or as a part of a complex system, is used to direct, by control of interactions with components of living systems, the course of any therapeutic or diagnostic procedure [1]. Biomaterials take the form of implants (dental implants, heart valves, intraocular lenses, ligaments, vascular grafts, etc.) or medical devices (artificial hearts, biosensors, pacemakers, etc.). They are used to replace or enhance the function of those tissues and organs that have been damaged or undergone degeneration. This allows for improving patients' quality of life.

The properties of biomaterials used in regenerative medicine largely depend on properties of the material used for their preparation (polymer, ceramic, metal, carbon, or their combination—a composite material); forms: solid material, porous material; spatial construction as well as physical and chemical properties.

BIOMATERIAL TYPES

Basically, biomaterials can be divided into two categories: (i) resorbable materials, which after implantation are getting disintegrated and resorbed in the tissue environment, and (ii) nonresorbable ones that will not get disintegrated and resorbed.

Biomaterials can be also as well classified according to multiple principles, such as bioinert and bioactive, biostable, biodegradable, etc. Active biomaterials encourage bonding with the surrounding tissue, while inactive biomaterials cause

Stem Cells and Biomaterials for Regenerative Medicine. https://doi.org/10.1016/B978-0-12-812258-7.00007-1

minimal or null tissue response. Degradable (resorbable) materials gradually release their mass into the surrounding tissue and may dissolve completely over a period of time. As a general rule, metals are inert; polymers may be inert or resorbable; and ceramics may be inert, resorbable, or active [2].

For any application, a given biomaterial must allow for sufficiently long-term use without rejection by the host organism, nor losing its intended function. To accomplish that, the design and selection of biomaterials must account for multiple properties. These properties can be divided into mechanical, physical, chemical, and biological, but in truth these categories are often intertwined. Biomaterials used for different purposes often have different requirements, so their ideal properties are always judged in their specific context.

Biomaterials must be biofunctional [3]. This means they must possess properties that make them able to carry out their specific function, either as implants or devices. It could be compensating for loss of function in diseased or damaged tissues, outright replacing the function of such tissues, or being used for diagnostic purposes. Examples of such properties, some of which are unique for their specific applications, are as follows:

- Space filling for cosmetic surgery [3].
- Load transmission and stress distribution for bone replacements (Fig. 7.1).
- Articulation for an articular cup of the hip joint.
- High lubricity for joint implants to counteract the mechanical wear of their use.
- Clarity and the correct refraction value (light-bending properties) for intraocular lens, to imitate light transmission properties of natural lens.
- Sound transmission for cochlear implants.
- Conductivity to electrical stimuli for pacemakers.
- Permeability for membranes, including dialysis membranes [3].

(A) (B)

FIG. 7.1 The structure of natural- (A), and synthetic hydroxyapatite (B) as observed using SEM (scanning electron microscope).

Biomaterials must be biocompatible. This means the host response to their implantation and presence cannot have adverse effects either on the implant, or on the host organism. Being safe for the host organism means they cannot cause harmful effects either locally, in the area of implantation, nor for other organs or tissues [3, 4]. Biomaterials must be:

- Blood compatible
- Nontoxic, unless it is their intended function, like in the example of "smart bomb" drug delivery systems designed to target and destroy cancer cells [3]
- Noninflammatory
- Nonpyrogenic
- Nonallergenic
- Noncarcinogenic.

Biomaterials must also account for the structure and function of the surrounding tissues and organs, so they do not disrupt them. Properties of an ideal biomaterial may change depending on the exact location of their implantation, and even on the patients' medical history [3]. Some biomaterials even incorporate living cells in order to enhance their biological and mechanical compatibility with the living tissue.

Biomaterials must be resistant to corrosion. This is a problem especially in the context of metallic implants. The presence of body fluids allows metallic implants to release metal ions that may accumulate in nearby tissues or be transported to other parts of the organism. Such ions may have adverse effect on the host's health, which makes them nonbiocompatible. Their release from the implant may also reduce the lifespan of the implant, hence, requiring a revision surgery. Notably, corrosion is not just dependent on chemical properties of the metal, but is also influenced by mechanical forces. Over the time of use, friction between the implant and other tissues may magnify the release of ions [3, 5, 6]. Interestingly, some biomaterials use an opposite approach and are intended to undergo degradation within the host's organism. However, even then it is important for the compounds that the released degradation product to be biocompatible.

MECHANICAL PROPERTIES

Strength

This value describes the magnitude of stress at which the material begins to fracture. When the bone implant interface starts to fail, damaged areas may be filled with soft, fibrous tissue, which allows more relative motion between the implant and the bone under loading [7]. Over time, this leads to increasing amounts of pain for the patient and requires a revision surgery [3]. The same implant may exhibit different strengths, depending on the nature and vector of forces applied to it. As a result, its strength is described through multiple values such as compressive strength, tensile strength, or yield strength.

Elastic Modulus, Also Known as Young's Modulus

This value describes the capacity of the biomaterial for elastic deformation under stress or its rigidity [8]. The desired value for a bone replacement must be as close to that of the connected bones as possible. A large difference in Young's modulus between the implant and surrounding bones can lead to load shielding from the natural bones, leading to their weakening and deterioration, which in turn leads to failure of the implant/bone interface [7, 9].

Hardness

Much like the elastic modulus, it should be as close as possible to bone hardness. If between a bone and a metal, one of them is softer than the other, it may end up penetrated by the harder material [8].

Fatigue Resistance

Fatigue characteristics strongly depend on microstructures of the biomaterial. Fatigue is caused by repeated cyclic loads that alter these microstructures over the time of use, causing fractures and deformations. This can lead to major problems associated with implant loosening, stress-shielding, and ultimately implant failure. Due to intense mechanical forces, this is frequently reported for hip prostheses [10]. In the case of metallic biomaterials, their initial microstructure is especially strongly determined by the type of processing and heat treatment used in their fabrication [4].

High Wear Resistance

Low wear resistance or a high friction coefficient may eventually result in loosening of the implant, or releasing debris into the surrounding tissues. The latter may cause a severe inflammatory response, which in turn can lead to destruction of surrounding bone [3, 11]. Susceptibility to wear also accelerates corrosion and releases nonbiocompatible metal ions from the implant [6].

Creep Resistance

(Cold flow resistance), which means low susceptibility to plastic deformation of material under constant pressure over time [12].

Tissues can be generally divided into hard tissues and soft tissues. The former include bone and tooth. The latter include blood vessels, cartilage, ligaments, skin, and others. Since biomaterials need to match the strength and modulus of surrounding tissues, metals and ceramics are used for hard tissue applications, while polymers are used for soft tissue applications [13].

OSSEOINTEGRATION

Osseointegration (osteointegration) refers to a phenomenon where an implant becomes so fused with bone that they cannot be separated without fracture [14]. When an implant is osseointegrated, there is no progressive relative movement between it and the bone with which it has direct contact. This means that osseointegration is necessary for long-term stability of prosthesis [15, 16]. Responsibility for osseointegrations rests mainly on living tissue through mechanisms that resemble direct fracture healing. The drilling of an implant cavity constitutes an injury to bony tissue leading to distinct phases of wound healing. The implant does not need to actively promote this state. Not causing a negative tissue response is enough to allow a degree of osseointegration [17–19]. However, the ultimate effectiveness of osseointegration does depend on surface topography of an implant. The properties of the implant's surface are crucial for adhesion and differentiation of osteoblasts during the initial phase of osseointegration as well as in long-term bone remodeling [20, 21]. These properties can be modified at microscale by such manufacturing techniques like acid etching, grit blasting, machining, and sandblasting [22]. These methods are intended to produce an optimal degree of roughness to promote osseointegration. Microscopic pits, grooves, and protrusions increase the surface area to act as a foothold to bone tissue and they set the stage for biological responses at the bone-to-implant interface [22, 23]. Surface topography matters also at nanoscale, where it interacts with tissues at cellular and protein level [24]. This involves not only aforementioned changes in surface roughness, but also surface chemistry [25]. These properties can be modified through discrete crystalline deposition (DCD), anodic oxidation, laser ablation, Titanium Oxide blasting, and acid etching [26].

SURFACE HYDROPHILICITY

Besides topography and roughness, hydrophilicity of implants is another key aspect of osseointegration. Hydrophilic surfaces are necessary so that proteins can maintain their conformation and function, whereas hydrophobic implant surfaces have been argued to cause changes in conformational [27].

PHOTOFUNCTIONALIZATION

UV treatment of implant surfaces enhances bioactivity and osseointegration. UV treatment reduces the degree of surface hydrocarbon and increases surface energy and hydrophilicity [26]. UV light has also been shown to restore bioactivity lost through age-related degradations [28].

SURFACE COATINGS

Surface coatings can be applied to implants to mimic biochemical and nanostructural properties of human bone. Such coatings include drug coatings, extracellular matrix proteins, growth factors, hydroxyapatite and nanocomposite coatings, messenger molecules, and peptides [26]. Biomaterial surface properties also determine implant's susceptibility to colonization by bacteria, which have a range of negative effects.

IMPLANTABLE MATERIALS

Metals and their alloys have been used almost exclusively for load-bearing implants, such as hip and knee prostheses and fracture fixation pins, plates, screws, and wires. The use of pure metals is rather uncommon, since alloys often exhibit improved material properties, such as strength and corrosion resistance. There are three material groups that dominate among biomedical metals:

Stainless Steels

Used primarily for their resistance to chemical corrosion [3, 29]. It is in use thanks to such advantages as high availability, relatively low cost, excellent fabrication properties, great strength, and accepted biocompatibility.

Cobalt-Chromium Alloys

Although pure Cobalt is not resistant to corrosion, Cobalt-chromium-based alloys are highly resistant to it, even in chloride-rich environment. This is thanks to presence of chromium, which leads to spontaneous formation of a layer of chemically inert chromium oxide (Cr_2O_3), when in contact with human body fluid environment [30].

Molybdenum and Nickel are often used to increase both strength and corrosion resistance of an alloy. Tungsten improves the strength of an alloy, but reduces its resistance to corrosion, which is particularly problematic in alloys containing Nickel, since corrosion releases toxic Ni^{2+} ions. As a result, alloys containing Tungsten are used only for short-term implants, such as bone plates and wire [30].

Properties of an alloy depend not only on its composition, but also its fabrication process. The F75-98 alloy is used in dentistry and is formed through mold casting. F799-99 has the same nominal composition, but is forged, which gives it a different classification [30].

Long-term clinical use has proven the high biocompatibility of CoCr-based alloys, but shown a certain susceptibility to mechanical damage and harmful biological reactions with Co, Cr, and Ni ions. Despite of these shortcomings, wrought CoCrMo alloys are the most popular metallic materials for permanent joint (bearing) site replacement [30].

Titanium Alloys

Titanium alloys exhibit three structural types: Alpha, Alpha-Beta (metastable-beta), and Beta. The Beta phase exhibits a lower modulus than Alpha, and meets most of other requirements for orthopedic application [31, 32], such as high strength, low density, high specific strength, good resistance to corrosion, complete inertness to body environment, enhanced biocompatibility, moderate elastic modulus (approximately 110 Gpa). On the other hand, long-term use of Titanium alloys shows release of Vanadium and Aluminum ions, both associated with long-term health problems, like Alzheimer's and neuropathy [7, 33]. Also Titanium suffers from extreme wear when rubbed against other metallic elements [34].

Ceramics

Ceramics are polycrystalline materials. They are very brittle—sensitive to the presence of cracks or other defects, and have poor resistance against tension forces. As a result, they are used less extensively than either metals or polymers. On the other hand, they exhibit hardness, great strength, stiffness, low density, and are resistant to wear and corrosion. Ceramics are usually electrical and thermal insulators. Ceramics are used in dentistry, orthopedics, and as medical sensors [3].

Bioinert ceramics retain their structure within the body after implantation and do not induce any immunologic host reactions. Bioactive ceramics form direct chemical bonds with bone or even with soft tissue of a living organism.

Alumina (Al_2O_3)

High-density high-purity (>99.5%) alumina was the first ceramic to find wide clinical use. It is used in load-bearing hip prostheses and dental implants, because of its combination of excellent corrosion resistance, high wear resistance, high strength, and good biocompatibility. Alumina's high resistance to wear stems from high surface energy and surface smoothness of this ceramic. The biocompatibility of alumina ceramic has been tested by multiple researchers, and it has shown no toxicity toward bone marrow cells [35]. Noiri et al. [36] implanted alumina-ceramic material within eye sockets of albino rabbits. After a period of 4 weeks, fibroblast proliferation and vascular invasion were observed and by eighth week, tissue growth was noted in the pores of the implant and no signs of either implant rejection or prolapse were observed [36].

Zirconia (ZrO_2)

Bioceramics made from Zirconia have advantages over those made of Alumina. These include higher fracture toughness, higher flexural strength, and lower Young's modulus. On the other hand, they have less favorable friction and wear properties, and when in contact with physiological fluids, their strength is reduced over time. In addition, natural ZrO_2 is often accompanied by Thorium and Uranium, which are difficult and expensive to remove. Both elements are radioactive and raise concern for mutagenesis of affected tissues [30].

Pyrolytic Carbon

Unlike metals, polymers, and other ceramics, these carbonaceous materials do not suffer from fatigue. However, their use is limited in load-bearing applications, because of their brittleness and low tensile strength. Investigation has shown that intrabone implantation results in mechanical bonding between the tissue and the implant after 3 months, but is accompanied with loss of implant strength [37].

Glass Ceramics

Glass ceramics may exhibit various degrees of bioactivity. The most bioactive glasses are composed of SiO_2, Na_2O, CaO, and P_2O_5. Properties of any type of glass depend on proportions of these compounds. It is possible to produce glass-based implants that are either nearly inert, resorbable, or bioactive (bioglasses). The exact parameters can be regulated by changing the composition of the material and thermal processing parameters for the ceramic [30]. Bioglasses has been widely used for filling bone defects, due to their tendency to bond with bone tissue. The porosity of bioglass enables increased resorption and bioactivity [2]. However, fabrication of glass ceramics faces the following hurdle. The heat treatment intended to improve their mechanical properties also causes their crystallization, which leads to reduced, or even nullified bioactivity. The 45S5 glass is an exception to this rule, as it retains bioactivity along its improved strength [30].

Bioresorbable Materials

Bioresorbable materials undergo gradual hydrolytic breakdown in the body while at the same time being replaced by natural tissue. The chemical byproducts of the degrading materials are incorporated into metabolic processes of the body and either absorbed into or released from the organism.

Calcium Phosphate

Calcium Phosphate ceramics come both as resorbable and bioactive materials. One of their main characteristics is their porosity. The ideal pore size for these bioceramics is close to that of spongy bone [3]. In order to bind with living bone, bioactive calcium phosphate materials must form a bone-like apatite layer on their surface [38].

POLYMERIC BIOMATERIALS

Poly(methyl Methacrylate)

PMMA can be prepared under ambient conditions by mixing powdered polymer with monomeric methyl methacrylate. The resulting "dough" is easy to use in the operating theater or dental clinic and explains its use in dentures and bone cement. The PMMA cement also assists in functioning of joint prostheses. However, it is a hard but brittle polymer, which makes it unsuitable for most clinical applications.

Ultra-High-Molecular-Weight Polyethylene (UHMWPE)

UHMWPE is used as the bearing surface in total joint arthroplasty; it has 90% success rate at 15 years with metal on polyethylene. Although in presence of polyethylene submicron particles are found in periprosthetic tissues, UHMWPE is the best material developed to date. The wear rates and coefficient of friction between polyethylene and cobalt-chromium are reported to be very low.

Other polymeric biomaterials include:

- Nylon 6/6
- Polyvinyl alcohol, Fig. 7.2 (PVA)

FIG. 7.2 The morphology of obtained polyvinyl alcohol (PVA) nanofibers, as seen under SEM.

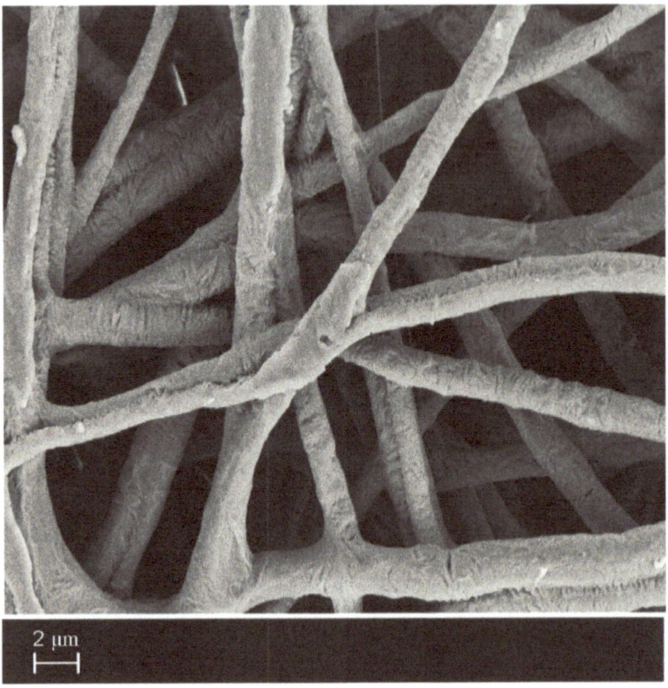

FIG. 7.3 The polycaprolactone (PCL) microfibers with 7% addition of poly lactide-*co*-glycolide (PLGA) as seen under SEM.

- Poly(ethylene terephthalate)
- Polycaprolactone, Figs. 7.3 and 7.4 (PCL)
- Poly(lactic acid)
- Polypropylene
- Polytetrafluoroethylene
- Silicone rubber
- Chitosan.

BIOCOMPOSITES

Biocomposites contain two or more constituent phases. This takes form of reinforcing materials suspended in a polymer matrix phase. Neither of these phases dissolve into each other nor merge with one another, but they still determine the overall properties of the implant. Varying the type and distribution of phases allows for an expanded range of biological and mechanical traits. The matrix can be made of a wide variety of polymers, both natural and synthetic. Reinforcing phases can be particulate matter (powders), short fibers and long fibers. The powders are represented by alumina, bioglass, carbon, hydroxyapatite, silica, titanium, and bone particles. The fibers may be made from carbon, glass, kevlar,

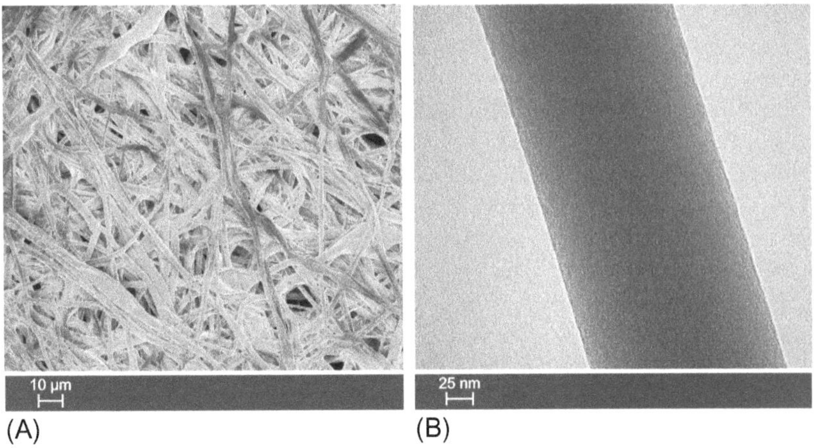

FIG. 7.4 The structure of porous polycaprolactone (PCL) fibers, as observed under SEM (A), and TEM (transmission electron microscope) (B).

or they may have organic origin, such as alginate, collagen, or gelatin [30]. Composites for internal prosthetic applications are most commonly composed of a polymeric matrix reinforced with chopped carbon fiber. They exhibit improved mechanical properties compared to constituent materials [39, 40]. In general, composite structures are typically produced from multiple laminate layers [41]. However, laminates are not currently in clinical use because there is no efficient method to modify the shapes of the implants intraoperatively to fit the bone. Biocomposites containing carbon fibers are not currently used for larger implants. The reasons for that are that they release carbon fibers into the surrounding tissues, and because it's difficult to predict the resorption of polymers in larger load-bearing implants.

BIOMATERIALS (SCAFFOLDING) IN TISSUE ENGINEERING

The goal of tissue engineering is to recreate natural tissues by using for this purpose mainly patient's own cells with resorbable and/or nonresorbable biomaterials. The natural ECM (Extra Cellular Matrix) [42], scaffold is an integral part of every tissue. ECM supports cells, gives the tissue elasticity, flexibility, mechanical strength, and other functional properties. The matrix is made of protein components (mainly collagen type I), polysaccharide (like hyaluronic acid), as well as sometimes, ceramic components (including hydroxyapatite, which is the basic building block of bone tissue, and teeth, Fig. 7.1).

Biomaterials are used in two ways [43]:

(i) Closed systems—in which the donor's cells are encapsulated in a protective and semipermeable membrane, then in created space cells make copies of themselves and release bioactive substances necessary for the body's functioning. The capsule also plays a protective role against the recipient's immune system response.

(ii) Open systems—in which surface of the obtained scaffolding has a direct contact with the patient's cells forming a supporting construction for growing cells. Scaffolds made of biodegradable materials such as PCL [44] (Fig. 7.4), PLA, chitosan [45], PVA (polyvinyl alcohol), PEO (poly(ethylene oxide)) [46] will resorb, while nonbiodegradable materials such as PE, titanium will be overgrown by the target tissue. Open systems are the most popular ways to use tissue scaffolds in tissue engineering. They are obtained with the use of: (a) incremental/additive technologies: FDM (fused filament fabrication), SLS (selective laser sintering), SLM (selective laser melting), bioplotter, (b) textile technologies including spunbond, meltblown, electrospinning [47] and (c) other technologies, and combined technologies, which typically represent the combinations of some of the mentioned technologies.

REFERENCES

[1] Biomaterials. Elsevier; 2017, https://www.journals.elsevier.com/biomaterials; (Accessed October 7, 2017).

[2] AZoM. Biomaterials—an overview, https://www.azom.com/article.aspx?ArticleID=108; 2001. (Accessed October 7, 2017).

[3] Patel Nitesh R, Gohil Piyush P. A review on biomaterials: scope, applications & human anatomy significance. Int J Emerg Technol Adv Eng 2012;2(4).

[4] Boretos JW, Eden M. Contemporary biomaterials, material and host response, clinical applications, new technology and legal aspects. Park Ridge, NJ: Noyes Publications; 1984. p. 232–3.

[5] Okazaki Y, Gotoh E. Comparison of metal release from various metallic biomaterials in vitro. Biomaterials 2005;26(1):11–21.

[6] Singh R, Dahotre NB. Corrosion degradation and prevention by surface modification of biometallic materials. J Mater Sci Mater Med 2007;18(5):725–51.

[7] Geetha M, et al. Ti based biomaterials, the ultimate choice for orthopaedic implants—a review. Prog Mater Sci 2008;54:397–425.

[8] Hartsuijker C, Welleman JW. Engineering mechanics. vol. 2. The Netherlands: Springer; 2001.

[9] Au AG, et al. Contribution of loading conditions and material properties to stress shielding near the tibial component of total knee replacements. J Biomech 2007;40(6):1410–6.

[10] Teoh SH. Fatigue of biomaterials: a review. Int J Fatigue 2000;22(10):825–37.

[11] Ramsden JJ, et al. The design and manufacture of biomedical surfaces. CIRP Ann Manuf Technol 2007;56(2):687–711.

[12] Ashby MF, Jones DRH. Engineering materials 1: An introduction to their properties and applications. Pergamon Press; 1980.

[13] Williams D. An introduction to medical and dental materials, concise encyclopedia of medical & dental materials. Oxford: Pergamon Press and The MIT Press; 1990. p. 17–20.

[14] Brånemark PI. Vital microscopy of bone marrow in rabbit. Scand J Clin Lab Invest 1959;11(Suppl 38):1–82.

[15] Brånemark PI. Osseointegration and its experimental studies. J Prosthet Dent 1983;50: 399–410.

[16] Rigo ECS, Boschi AO, Yoshimoto M, Allegrini SJ, Konig BJ, Carbonari MJ. Evaluation in vitro and in vivo of biomimetic hydroxyapatite coated on titanium dental implants. Mater Sci Eng C 2004;24:647–51.
[17] Linder L, Albrektsson T, Brånemark PI, Hansson HA, Ivarsson B, Jönsson U, Lundström I. Electron microscopic analysis of the bone-titanium interface. Acta Orthop Scand 1983;54: 45–52.
[18] Stanford CM, Keller JC. The concept of osseointegration and bone matrix expression. Crit Rev Oral Biol Med 1991;2:83–101.
[19] von Wilmowsky C, Moest T, Nkenke N, Stelzle F, Schlegel KA. Implants in bone: Part I: A current overview about tissue response, surface modifications and future perspectives. Oral Maxillofac Surg 2014;18(3):243–57.
[20] Junker R, Dimakis A, Thoneick M, Jansen JA. Effects of implant surface coatings and composition on bone integration: a systematic review. Clin Oral Implants Res 2009; 20(Suppl 4):185–206.
[21] Le Guéhennec L, Soueidan A, Layrolle P, Amouriq Y. Surface treatments of titanium dental implants for rapid osseointegration. Dent Mater 2007;23(7):844–54.
[22] Dohan Ehrenfest DM, Coelho PG, Kang B-S, Sul Y-T, Albrektsson T. Classification of osseointegrated implant surfaces: materials, chemistry and topography. Trends Biotechnol 2010;28(4):198–206.
[23] Fischer K, Stenberg T. Prospective 10-year cohort study based on a randomized controlled trial (RCT) on implant-supported full-arch maxillary prostheses. Part 1: Sandblasted and acid-etched implants and mucosal tissue. Clin Implant Dent Relat Res 2012;14(6):808–15.
[24] Mendonça G, Mendonça DBS, Aragão FJL, Cooper LF. Advancing dental implant surface technology—from micron- to nanotopography. Biomaterials 2008;29(28):3822–35.
[25] Coelho PG, Jimbo R, Tovar N, Bonfante EA. Osseointegration: hierarchical designing encompassing the macrometer, micrometer, and nanometer length scales. Dent Mater 2015;31(1): 37–52.
[26] Ralf S, Berndt S, Frank S, Benedicta B-B, et al. Impact of dental implant surface modifications on osseointegration. Biomed Res Int 2016;2016:16.
[27] Terheyden H, Lang NP, Bierbaum S, Stadlinger B. Osseointegration—communication of cells. Clin Oral Implants Res 2012;23(10):1127–35.
[28] Hori N, Ueno T, Suzuki T, et al. Ultraviolet light treatment for the restoration of age-related degradation of titanium bioactivity. Int J Oral Maxillofac Implants 2010;25(1):49–62.
[29] Navarro M, et al. Biomaterials in orthopaedics. J R Soc Interface 2008;5(27):1137–58.
[30] Qizhi C, George T. Biomaterials: a basic introduction. Boca Raton: CRC Press, Taylor & Francis Group; 2015.
[31] Eisenbarth E, et al. Biocompatibility of â-stabilizing elements of titanium alloys. Biomaterials 2004;25(26):5705–13.
[32] Nag S, et al. Characterization of novel borides in Ti–Nb–Zr–Ta+2B metal-matrix composites. Mater Charact 2009;60(2):106–13.
[33] Navarro M, et al. Biomaterials in orthopaedics. J R Soc Interface 2008;5(27):1137–58.
[34] Miller PD, Holladay JW. Friction and wear properties of titanium. Wear 1958;2(2):133–40.
[35] Kanematsu N, Shibata K, Yamagami A, Kotera S, Yoshihara Y. Shika Kiso Igakkai Zasshi 1985;27(1):382–4.
[36] Noiri A, Hoshi F, Murakami H, Sato K, Kawai SI, Kawai K. Biocompatibility of a mobile alumina-ceramic orbital implant. Folia Ophthalmologica Japonica 2002;53(6):476–80.
[37] Lewandowska-Szumieł M, Komender J, Gorecki A, Kowalski M. Fixation of carbon fibre-reinforced carbon composite implanted into bone. J Mater Sci Mater Med 1997;8(8):485–8.

[38] Thian ES, Loh NH, Khor KA, Tor SB. In vitro behavior of sintered powder injection molded Ti-6Al-4V/HA. J Biomed Mater Res 2002;63(2):79–87.
[39] Christensen R. Mechanics of composite materials. New York: Wiley; 1979.
[40] Cogswell FN. Thermoplastic aromatic polymer composites. UK: Butterworth Heinemann; 1992.
[41] Beatty MW, Swartz ML, Moore BK, et al. Effect of microfiller fraction and silane treatment on resin composite properties. J Biomed Mater Res 1998;40:12–23.
[42] Davies JA. Extracellular matrix. In: Encyclopedia of life sciences. Chichester: John Wiley & Sons, Ltd; 2001. https://doi.org/10.1038/npg.els.0001274.
[43] Estima Gomes MM. A bone tissue engineering strategy based on starch scaffoldsand bone marrow cells cultured in a flow perfusion bioreactor. Outubro: Universidade Do Minho Escola De Engenharia Grupo De Investigação 3b's—Biomateriais, Biodegradáveis, Biomiméticos Departamento De Engenharia De Polímeros; 2004.
[44] Hudecki A, Gola J, Ghavami S, Skonieczna M, Markowski J, Likus W, Lewandowska M, Maziarz W, Los MJ. Structure and properties of slow-resorbing nanofibers obtained by (co-axial) electrospinning as tissue scaffolds in regenerative medicine. PeerJ 2017;5:e4125.
[45] Pillai CKS, Paul W, Sharma CP. Chitin and chitosan polymers: chemistry, solubility and fiber formation. Prog Polym Sci 2009;34:641–78.
[46] Pakravan M, Heuzey M-C, Ajji A. Core-shell structured PEO-chitosan nanofibers by coaxial electrospinning. Biomacromolecules 2012;13:412–21.
[47] Meng ZX, Wang YS, Ma C, Zheng W, Li L, Zheng YF. Electrospinning of PLGA/gelatin randomly-oriented and aligned nanofibers as potential scaffold in tissue engineering. Mater Sci Eng C 2010;30:1204–10. https://doi.org/10.1016/j.msec.2010.06.018.

Chapter 8

Overview of Technologies Applied for Biomaterial Production

Chapter 8.1

Overview of Technologies Applied for Biomaterial Production—3D Printing

Wojciech Gaweł*, Frank Schweizer†
*Solveere Sp z o.o., Ogrodzieniec, Poland
†Department of Chemistry, University of Manitoba, Winnipeg, MB, Canada

INTRODUCTION

Additive manufacturing (AM) technologies have been under development since early 1980s, and they have found its place in the modern world of manufacturing, science, and even entertainment. Success of AM technologies can be found in its foundations making possible to achieve things for a long time seen only in sci-fi movies. Building complex models in a fraction of time needed by using standard technologies makes it possible to build products in units rather than thousands and allows to move the place of production from factories to offices and homes and finally providing wide range of tools to medicine and making personalized treatment more achievable. Most important feature of AM technologies is the way of building 3D objects by adding material [1] rather than removing it like in CNC (computerized numerical control) technologies. This approach of manufacturing provides advantages over any other technology in the field. For instance, imagine the bone with lots of cavities and porosity structures. Geometry, like this, is impossible to be cut out of block of material. It is also not possible to make it using injection molding or casting of any sort. The only way to produce a bone besides growing is to print it in 3D. AM also

Stem Cells and Biomaterials for Regenerative Medicine. https://doi.org/10.1016/B978-0-12-812258-7.00008-3

does not need any tools for fabricating parts, nor complicated programming procedures. Most of systems are automatized and do not need the supervisory of human. You may wonder why we are not using the AM technologies for everything if they are so great. Truth is that AM also has certain disadvantages and there are many of them.

(i) The first disadvantage is the cost of manufacture. Each technology has its own cost, but even the cheaper one will never be competitive to injection molding when it is about making large-scale production.
(ii) The second disadvantage is connected to the first one and its time of production. It will never be faster than traditional ways of manufacturing when making large-scale production.
(iii) The third disadvantage is caused by the AM technology characteristics and it's the final parts strength. Most of the technologies are showing much less resistance in the axe of building the model. Arrangement of the part will have impact of its durability, so it's important to prepare a setup with knowledge of final model's application. Only few technologies like DMLS (Direct Metal Laser Sintering)/SLM (Selective Laser Melting) do not show much difference depending on the printed parts orientation.
(iv) Finally, the last disadvantage, I would like to highlight is the limited number of materials we can use for 3D printing. Not all materials available for wide range of production will be good for Additive Manufacturing. Luckily, this number is growing constantly.

Despite the drawbacks listed, additive manufacturing remains the technology, which is expected to take us to the future thanks to its organic more than fabric way of production. Cheap 3D printers and widely available rapid prototyping services providers are shortening the distance between people with idea and others, willing to buy it as materialized object: R&D center in Your computer and factory on Your desk.

Additive manufacturing is also perfect tool for medicine. Drug development, tissue engineering, prosthetics, implantation, tools designing, and manufacturing—those are the fields on which 3D printers and Additive Manufacturing have now much to offer. There are many ways of classification we may apply for technologies, but for simplicity I have chosen to make the division based on materials used for printing parts.

UV AND LIGHT-SENSITIVE RESINS

One way of building three-dimensional models is by using resins sensitive to light. Usually, UV-light is applied to cure (harden) materials. Laser, UV lamp, or projector could be used as light source. Each approach has its advantages and drawbacks. In general, resin-based AM technologies give the best accuracy and best surface finish; however, resin-based parts are not heat resistant, not particularly hard, and they are not stable for longer use in a sun-exposed environment. Some novel resins have recently been developed that stay stable when exposed to UV light (sun-light) and some are also resistant to high temperatures. Good examples of using resin-based manufacturing are dental—as well as hearing-aid industries. In both cases, advantages of liquid resin-based manufacturing, like very good surface finish and high accuracy, are being appreciated. DLP (Digital Light Processing) technology uses projector for curing UV-sensitive resin. The main advantage of this approach is a very fast printing process. Whole layer is being cured in the same time, so there is no difference if we are printing only one model or full chamber, printing time will be same (Fig. 8.1.1).

FIG. 8.1.1 Flagship model of Envisiontec's DLP family—Envisiontec Perfactory 4 LED.

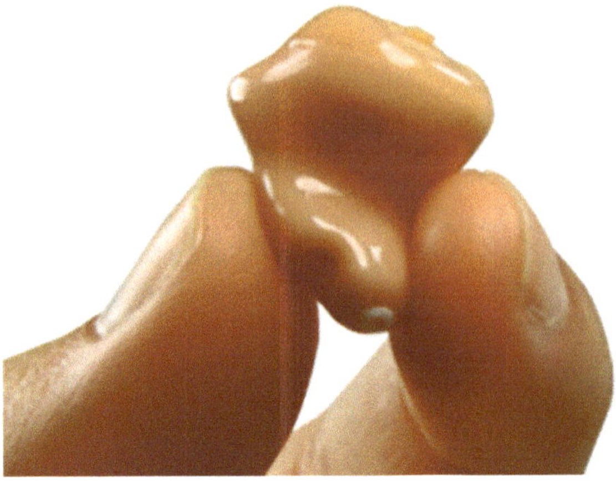

FIG. 8.1.2 3D printed hearing air using Envisiontec's DLP technology.

A leading manufacturer using this technology is Envisiontec GmbH. Envisiontec describes their technology as "industries that demand the highest part accuracy and smooth organic shapes, such as the hearing aid and dental industries, prefer EnvisionTEC 3D printers for mass customized production. Six out of ten hearing aids made globally today are made on EnvisionTEC printers, and dentists, orthodontists and dental labs are rapidly adopting EnvisionTEC's advanced DLP printers for the same reasons: highly accurate, smooth fuss-free surfaces." EnvisionTEC uses several patented and proprietary approaches to delivering these outstanding curves and smooth surface finishes (Fig. 8.1.2).

To understand why these methods work so well, let's first identify the root of the problem with a generic DLP approach. A projector delivers light in square-shaped pixels, which create volumetric pixels, also known as voxels, when the light cures resin. Along a curved edge, this causes what the industry refers to as "stairstepping"—a jagged edge reminiscent of an early-era 8-bit video game. Adding to the challenge, the size of each pixel is predetermined by an easy formula: each pixel is the size of the build space, or work envelope, divided by the resolution of the projector. The smaller the object being printed, the less of an issue a user will have with stairstepping, but the bigger an item gets, the more pronounced the pixels and stairstepping gets. What's more important is the size of the pixels cannot be changed without changing the size of the projector or the build space. In the early 2000s, a small team of engineers in Germany and the United States led by Al Siblani, EnvisionTEC's founder, and Alexandr Shkolnik, EnvisionTEC's chief technology officer, developed two separate methods of smoothing out pixelated stairsteps. Their first solution, quite simply, cuts the size of each stairstep by half a pixel. EnvisionTEC filed a patent for this approach (US7790093 B2) in 2005. EnvisionTEC's Enhanced Resolution Module, or ERM, is a physical device—manufacturing in the company's Germany facility—that deliberately causes a tiny and precise amount of

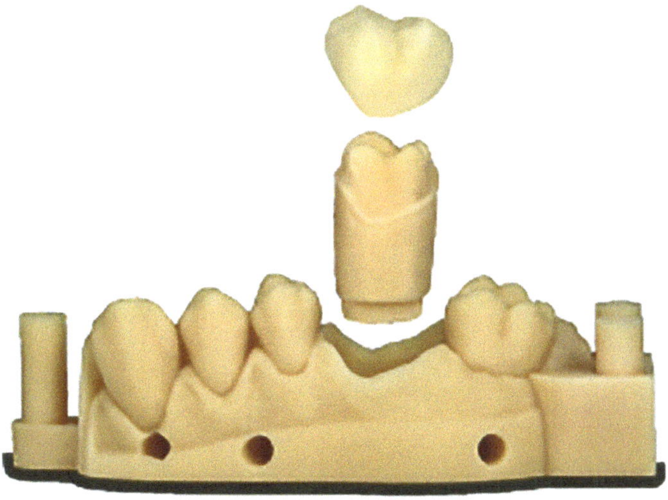

FIG. 8.1.3 Example of dental model printed with Envisiontec's DLP technology.

"pixel shift" in the projector delivering the light. This shift literally cuts the issue of stairstepping in half, for objects that are both large and small (Fig. 8.1.3).

This process is also why the curved surface finish from EnvisionTEC DLP printers is at least twice as good as its competitors, even when using the same resolution projectors. EnvisionTEC's Desktop and Perfactory DLP machines come with HD projectors in varying resolutions, from 1400×1080 to 1920×1200. The ERM device isn't a simple "module." Rather, it's a complete electromechanical system. A controller box delivers an electric current—one for movement in X and another for movement in Y—to a small square-shaped metal plate with a hollow interior that physically sits just below the projector. This device, when strategically excited by electricity, allows for a tiny amount of tightly controlled motion in X and Y.

The ERM literally shifts the projection a half pixel to the left or right in X or a half pixel up or down in Y. With generic DLP [2] technology, a 3D printer might flash a complete image layer for, say, 10s as it builds an object. With EnvisionTEC's ERM pixel-shifting technique, each exposure is done twice. In this example, let's say the first baseline exposure is for 5s where original curing should take place. But the next 5s exposure is shifted a half-pixel left in X and down in Y, reducing any stairstepping in half. Depending on the size of the final object being built, this reduction of stairstepping can make stairstepping virtually undetectable to the human eye. This pixel-shifting is completely controlled and regulated by EnvisionTEC's proprietary Perfactory software, too, so users do not need to manually control it. The system works automatically to smooth surface finish. But EnvisionTEC did not stop there in its quest to deliver the best surface finish. EnvisionTEC combines its pixel-shift strategy with another patented and proprietary method called grayscaling, which is, simply

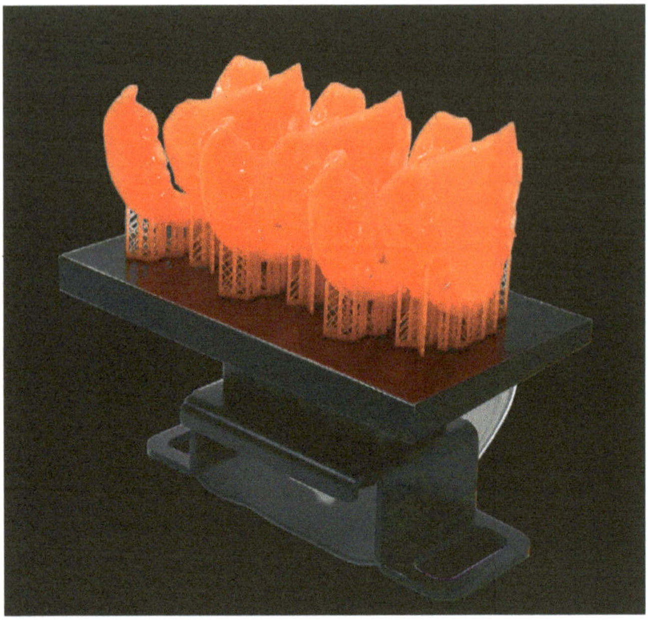

FIG. 8.1.4 Example of one 3D building job. All models have been printed in the same time with speed up to 25 mm/h in Z axis.

put, an antialiasing technology for 3D objects. The first patents for this method were filed in 2007, in both Germany and the United States (EP1849587 B1, US20080038396A1, US20070260349A1). Antialiasing technologies are prolific today—used in 2D digital platforms virtually everywhere images appear digitally, from video games and TV screens to pretty much anything that has a digital monitor. In today's digital world, there are many high-tech methods of antialiasing to smooth the edges of all the pixelated images we see every day (Fig. 8.1.4).

EnvisionTEC was the first 3D printing company to figure out a way to convert this antialiasing strategy to 3D voxels from a projector. So, after EnvisionTEC's ERM technology has cut the stairsteps in half, the company's proprietary gray-scaling technique smooths out whatever remains of these "jaggies," as stairstep images are known in video game circles.

Unlike an SLA laser beam that is either on or off, an HD light projector can emit light in many shades, or intensities, of light along a spectrum that goes from white and gray to black. In fact, there are 1024 discernable shades of gray between white and black. EnvisionTEC technology uses 255 of these shades, with pure white set at 255 to dark black set at zero, to feather out edges in a sophisticated and controlled fashion. EnvisionTEC's smart and proprietary software controls each individual pixel at the curved edge of an object, setting it in a feathered rainbow of grays, in order to deliver a smooth, antialiased edge.

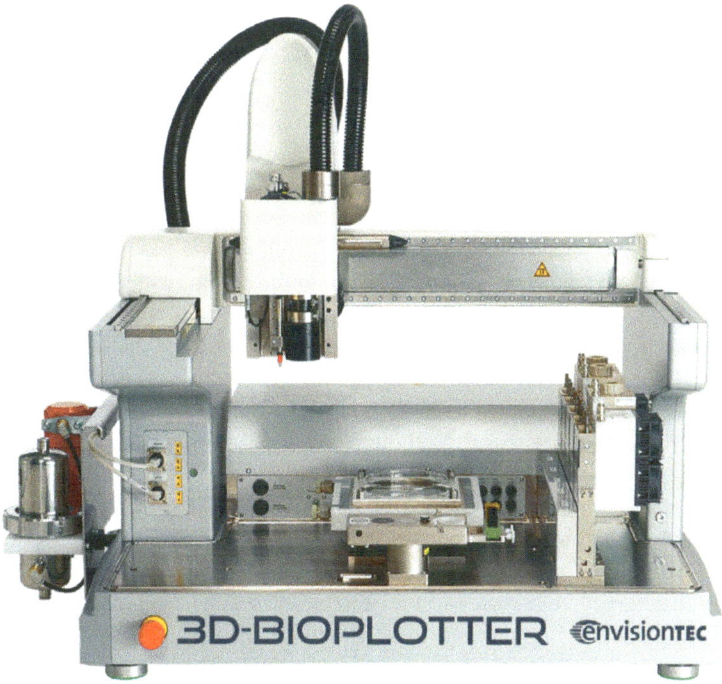

FIG. 8.1.5 Envisiontec's manufacturer series bioplotter.

What is even more fascinating is the varying intensities made possible with grayscaling and each exposure also delivers another unique effect with EnvisionTEC's DLP technology. While the X and Y dimensions of a pixel are fixed, the Z-depth of a voxel is adjustable depending on the intensity of light delivered to each pixel through grayscaling. That means while generic forms of DLP print flat layer by flat layer, EnvisionTEC DLP prints each exposure in varying depths—a layerless 3D printing strategy that enhances the smooth and accurate surface finish. The harmonious interplay of EnvisionTEC's pixel-shifting ERM technology, combined with grayscaling, delivers a market-leading surface finish with no striations or stairstepping. Scans of the final parts using this technology show that this technology is more accurate than SLA (laser-based stereolithography) (Fig. 8.1.5) [3, 4].

Envisiontec's approach for 3D printing allows fast and accurate printing. Best results are visible for small- and medium-sized models up to 200 × 200 mm. The limitation of DLP technologies is size of build platform. The bigger it is—the less accuracy could be gained. Advantages and limitation of this technology make it a perfect match for Dental and Hearing aid industry as most of printed models are within 10 × 10 cm in size.

POLYMER AND METAL POWDERS: SLS/SLM TECHNOLOGY

Another way of producing 3D models is SLS (selective laser sintering) [5] and SLM (selective laser melting) [6] technologies. While both are considered as different technologies, both share a lot of similarities in the approach of building parts. Both technologies are using material in the form of very fine powder with particles size not exceeding 60 microns. In addition, both are using laser to melt surface in selective areas and mechanical recoater for adding powder after each printed layer. Moreover, both are also using neutral gases in the building chamber to make the process more stable or in the case of using reactive metals to prevent explosion. However, the main difference between these technologies is the type of material used. Besides the material, there are many more differences, which prevent the machines to be converted from one technology to another. Nevertheless, materials to be used remain the main one visible for the first sight. Similarities are also in the limitations of shape that can be printed. We can use the general truth that if the model is easier to be cut or be done on CNC machine, then its harder to be made on SLS/SLM machine. On the other side, very complex shapes with cavities inside or openwork structures are not very demanding. Limitations of shapes make both technologies perfect for printing organic structures like bones. SLM technology is being used for manufacturing prosthetic parts and implants made of Titanium. Openwork structure and high roughness possible to gain in some areas of models are giving advantages in producing implants that are being overgrown by bone tissue. As I mentioned before, main difference is materials that we can use.

In SLM technology, we have a wide selection of metal powders available we can use. Existing technology makes it relatively easy to fabricate powder from metals, so it is possible to create material that will suit our demands. SLS technology uses polymers. The most popular materials are based on PA (polyamide) powder with very often filling of particles changing properties of PA itself to adjust performance we most care of. Good example of add-ons to the powder base is particles of glass fiber or aluminum. In both cases, it changes basic material characteristic to better suit desired applications. Advantages of SLS printing technology make it potentially applicable for the preparation of bone prosthetics or implants. However, currently we lack available material that would ideally suit this application. Given the intensive research in the area, SLS' full adaptation for regenerative medicine applications is expected within the following 5–10 years. For now, we may use SLS technology to print models from RTG or MRI files. Such models may help to better appreciate the changes with bone structures or could be used for operation planning. Printed bones even if made of PA resemble original bones much better than when prepared with any other 3D printing technology, but they are not yet suitable for implantation. Much closer to that goal is SLM technology, which already has been used and still is present in medicine. Implants printed with SLM technology are more accurate and better personalized than fabricated with traditional technologies.

Possible shapes and openwork structures give better options of connecting bones with implants as well as easier overgrowing by tissue after implantation.

BIOLOGICAL MATERIALS: BIOPRINTING

Bioplotter is a 3D printer capable of using biological material as "ink." Such machines are built just for one purpose—to produce models with highest accuracy with as many different materials as possible. This approach makes it perfect for researchers and developers (Fig. 8.1.6).

The leading company producing bioplotters is Envisiontec GmbH. The top model from bioplotter family enables to use five different materials in the same print process. It can use thermoplastics as well as gel or liquid materials. It may possess a heated or chilled printing platform, and chamber. Accuracy of positioning is at the level of 1 micron. The system uses modular components, such as sterilized heating and cooling cartridges, standard Luer-Lock syringes with standard needle-tip sizes, and also an easy-to-use 365-nm UV curing head.

Attempts have been made to print functioning organs while using bioplotters. While some of them, i.e., kidney would have the right shape color and consistency, they are still far away from achieving functionality. While bioplotters will certainly enter the field of clinical regenerative medicine, most likely, their initial applications will be to produce implantable elements for, i.e., pancreatic islets, regenerative heart patches, soft elements of joints, implants for cosmetic surgery, and implantable drug delivery systems or reservoirs.

A promising close-to-the-market application is to print implantable drug reservoirs, or "intelligent pills" for continuous drug release. For example, a 3D printed a humic acid-polyquaternium 10 (HA-PQ10) controlled release fixed dose combination (FDC) tablet comprising the anti-HIV-1 drugs, efavirenz

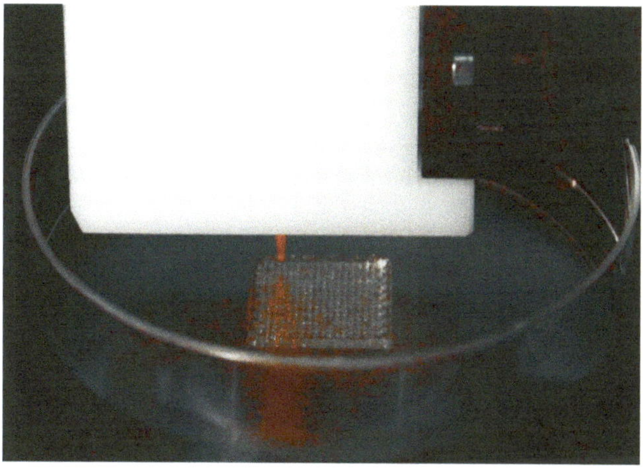

FIG. 8.1.6 Example of 3D printed scaffold using bioplotter.

(EFV), tenofovir disoproxil fumarate (TDF) and emtricitabine (FTC) has recently been described [7]. In vitro drug release studies were conducted in biorelevant media followed by in vivo study in the large white pigs, in comparison with a market formulation, Atripla®. The overall surface of the tablet exhibited a fibrilla structure and the 90° inner pattern was determined to be optimal for the release of drug combinations at a fixed dose.

These results demonstrate that intestinal-targeted and controlled drug release could be achieved, and also illustrated how 3D-printing could be used to generate pills where not only the shape but also the geometry of inner distribution of drugs affects their bioavailability over time. The HA-PQ10 drug entrapment technology achieved an enhanced relative bioavailability of EFV, TDF, and FTC compared to the market formulation for potentially enhanced HIV treatment.

The application of bioplotters in tissue engineering is quickly expanding. In the following experiment, bioplotter was used to generate a circular frame composed of polycaprolactone (PCL) biomaterial to which an acellular natural scaffold was attached and implanted subcutaneously in Sprague Dawley rats, for biocompatibility testing [8]. After 1 and 4 weeks, tissue scaffolds were explanted and evaluated by hematoxylin and eosin (H&E), Masson's trichrome, Picrosirius Red, transmission electron microscopy (TEM), immunohistochemistry, and mechanical testing.

As yet another example of the potential use for bioplotters are the attempts to produce artificial bone implants [9]. To meet the growing demand for bone tissue engineering to repair and regenerate parts of bones damaged by, i.e., trauma or disease, a series of attempts have been made to produce bone-replacement elements of defined shape, while using various materials. For example, a 3D printing technology, as an innovative fabricating tool, has been tested to generate a polymeric, bone-replacement scaffold fulfilling defined parameters [9]. For that purpose, sodium alginate as a bioink was applied, which interacted with a poly(ethyleneimine) solution during the bioprinting process, to form a polyelectrolyte complex through ionic bond formation. Silica gel was included in the bioink as temporal inorganic support component and for the enhancement of mesenchymal cell differentiation toward osteoblasts. The scaffold maintained its 3D architecture for at least 28 days, under conditions mimicking tissue environment. The scaffold's porosity was ~60%, and pore size was ~360 μm, respectively. Its Young's modulus of 18.37 MPa highlighting that the scaffold possesses the mechanical capabilities for certain bone-tissue-engineering applications. Besides sodium alginate and silica, other materials are being currently tested as ink's components of the 3D-printing of defined bone implants.

ACKNOWLEDGMENTS

Thanks to the courtesy of Envisiontec, I am able to share the examples of successful projects realized with the help of bioplotter (I, II, III).

REFERENCES

[1] Ngo TD, Kashani A, Imbalzano G, Nguyen KTQ, Hui D. Additive manufacturing (3D printing): a review of materials, methods, applications and challenges. Compos Part B Eng 2018;143:172–96.

[2] He R, Liu W, Wu Z, An D, Huang M, Wu H, Jiang Q, Ji X, Wu S, Xie Z. Fabrication of complex-shaped zirconia ceramic parts via a DLP-stereolithography-based 3D printing method. Ceram Int 2018;44(3):3412–6.

[3] Wanga J, Goyanesab A, Gaisforda S, Basita AW. Stereolithographic (SLA) 3D printing of oral modified-release dosage forms. Int J Pharm 2016;503(1–2):207–12.

[4] White Papers Envisiontec. Advanced DLP For Superior 3D Printing Learn how pixel-shifting & grayscaling technologies harmoniously anti-alias 3D edge surfaces to allow fast DLP printing with superior accuracy & surface finish. 3D Printing Materials by Envisiontec https://envisiontec.com/3d-printing-materials/accessed on October 4, 2018.

[5] Finaa F, Goyanes A, Gaisford S, Basit AW. Selective laser sintering (SLS) 3D printing of medicines. Int J Pharm 2017;529(1–2):285–93.

[6] Wei C, Li L, Zhang X, Chueh Y-H. 3D printing of multiple metallic materials via modified selective laser melting. CIRP Ann 2018;67(1):245–8.

[7] Siyawamwaya M, du Toit LC, Kumar P, Choonara Y, Kondiah PPPD, Pillay V. 3D printed, controlled release, tritherapeutic tablet matrix for advanced anti-HIV-1 drug delivery. Eur J Pharm Biopharm 2018. [in press]. Available online 12 April 2018. https://doi.org/10.1016/j.ejpb.2018.04.007.

[8] Khorramirouz R, Go JL, Noble C, Jana S, Maxson E, Lerman A, Young MD. A novel surgical technique for a rat subcutaneous implantation of a tissue engineered scaffold. Acta Histochem 2018;120:282–91.

[9] Sithole MN, Kumar P, du Toit LC, Marimuthu T, Choonara Y, Pillay V. A 3D bioprinted in situ conjugated co-fabricated scaffold for potential bone tissue engineering applications. J Biomed Mater Res A 2018;106:1311–21.

Chapter 8.2

Overview of Technologies Applied for Biomaterial Production—Electrospinning

Andrzej Hudecki*, Magdalena Wierzbik-Strońska†

*Institute of Non-Ferrous Metals, Gliwice, Poland

†Katowice School of Technology, The University of Science and Art, Katowice, Poland

INTRODUCTION

In order to conclusively evaluate the usefulness of microfibers and nanofibers for medical purposes, it is necessary to take into account, first of all the influence of: (i) process conditions, (ii) solvent properties, as well as (iii) environmental conditions, which as a whole influence the structure and properties of the obtained fibers. As of the present, polymer nanofibers are no longer just applied in the textile industry but are developing in fields connected with medicine [1], filters [2], sensors [3], the military, pharmacology, the so-called flexible electronics [4], photovoltaics [5], environmental protection [6], as well as composite materials [7].

Several methods of obtaining nanofibers exist:

(i) *Drawing* [8] a method, which enables obtaining nanofibers by immersing a micropipette in a selected solution, then removing it with a speed of 100 µm/s in order to obtain the fiber; due to low efficiency, the method did not find any application in industry.

(ii) *Template synthesis*, a method, which is based on using membranes with porosity of somewhere between several tens to several hundred nanometers made of aluminum oxide (Al_2O_3) or silicon monoxide (SiO_2), which are used as a template to obtain nanofibers, nanowires, or nanotubes [9].

(iii) *Phase separation*, a method based on the phenomenon of separating a multicomponent system into two phases: one enriched by a given component and the other lacking this component. The benefit of this method is high repetitiveness of the obtained structures [10].

(iv) *Molecular self-assembly*, a process due to the influence of intermolecular forces such as electrostatic force (Coulomb's law), hydrogen bonds, force of the stable dipole—stable dipole type, as well as the van der Waals forces. The mentioned forces are responsible for the phenomenon of forming orderly structures [11].

(v) Despite a variety of methods to obtain micro- and nanofibers, the most effective way to obtain them at the present is the process of electrospinning, i.e., stretching solution streams (*solution electrospinning* method) or melting (*melt electrospinning* method) under the influence of an electrostatic field. As a result of the impact of an electrostatic field, the following occurs: (a) solvent evaporation, if it is a method based on solutions (*solution electrospinning)* or (b) solidification of the jet of the polymer material, if it is a method based on previous melting of the polymer material, then stretching it in the electrostatic field, meaning *melt electrospinning*.

CONDITIONS AFFECTING THE ELECTROSPINNING PROCESS

The influence of factors on the structure and properties of the obtained fibers was described based on *solution electrospinning*, in this process the geometric characteristics and properties of the obtained fibers depend upon:

(i) Environmental factors, including:
  (a) Gas temperature
  The temperature of the gas in which the process of solution electrospinning takes place has a significant influence on the speed of the evaporation of solvents used in order to transform the polymer material into a solution; the higher the gas temperature the quicker the solvents will evaporate. A good example of a solvent, which evaporates too rapidly, is acetone. As research has shown [12], the speed of acetone evaporation contributed to the clogging up of the nozzles, which is an adverse phenomenon influencing destabilization of the whole process.
  (b) Humidity
  If air forms, the gas atmosphere where solvent evaporation occurs, the molecules of water vapor present in it, could in a significant way influence the stability of the process. Excess water molecules present in air will contribute to dissipate charges running through the stretched stream, which may destabilize the whole process.
  (c) Flow of gas
  It is necessary for the solvent steam released from the solution stream to be dissipated; on the other hand, the flow of gas should not influence the electrospinning process, which is sensitive to gas movement.
(ii) Process conditions, including:
  (a) Type of positive electrode
  The process of transforming the polymer solution stream into micro or nanofibers begins with a positive electrode. Depending on the type of the electrode, we can obtain varied fibers. In

a standard process of electrospinning, the positive electrode is a nozzle, or more appropriately a hollow metal needle, inside which the selected solution flows. When using this type of nozzle, we obtain whole cross-section micro- and nanofibers. If in the electrospinning process several or several tens of nozzles are used at the same time, we are dealing with a process of multi-jet electrospinning [13]. On the other hand in the process of coaxial electrospinning [14], the positive electrode is a nozzle made of two parts, the so-called nozzle in a nozzle in which we can differentiate: nozzle with a smaller diameter, the inner one through which solution "A" flows, and which is surrounded by an outer nozzle with a larger diameter than the inner nozzle, and through which solution "B" flows. These types of nozzles make it possible to obtain core-shell nanofibers [15, 16] as well as hollow nanofibers [17]. Nozzles used in the coaxial electrospinning process make it possible to combine various solutions with one another, which makes the micro- and nanofibers obtained in this way very attractive for application in regenerative and reconstructive medicine due to the possibility of placing in the fiber core, a biologically active substances that could be released in a controlled way, namely (i) only after the breakup of the outer shell, or (ii) released along with the breakup of the shell, (iii) or released with the breakup of the fiber core. In comparison with standard electrospinning, the technology of coaxial electrospinning enables: (1) insulating materials sensitive to atmospheric factors in a protective shell, (2) closing (encapsulating) medications, (3) obtaining new composite materials as a strengthening phase in these materials, as a result of combining a material with high resistance of the core with a material with good adhesion in contact with the matrix, (4) obtaining 3D scaffolds for tissue engineering, (5) transforming materials, which could not be transformed in a standard process of electrospinning into nanofibers, nanowires, or nanotubes. The nozzles used in the coaxial electrospinning process are used both in a method based on solvents, as well as in a method based on melting polymer materials.

In the mentioned standard and coaxial electrospinning processes, the solution flows through the nozzle interior, while the outer part of the nozzles is surrounded by gas. In the process of *bubble electrospinning* this dependency is reversed, meaning the nozzle is surrounded by a selected solution, while a gas flows through its interior, creating gas bubbles on the surface of the solution and then fibers are cut out from the bubble surface in direction of the second electrode [18].

(b) Types of negative electrodes

Spatial placement of the fibers depends on the type of negative electrode. Using flat collectors enables us to obtain fibers placed in a random way, i.e., in an isotropic orientation. On the other hand, using rotational collectors enables us to obtain aligned fibers, i.e., of anisotropic orientation. Spatial placement of fibers influences, among others, resistance of the obtained mats, for fibers placed in an isotropic way the resistance will be similar in all directions, while for fibers placed in an anisotropic way the resistance will vary depending on the spatial placement of the fibers.

(c) Solution flow
In the electrospinning process based on using nozzles, the speed of the solution flowing through the nozzle influences, among others, the thickness of the obtained fibers, the higher the speed, the larger the fiber diameter.

(d) Tension
In the electrospinning process, the solution flowing through the nozzle is subject to the impact of a strong electrostatic field. The electrostatic field is a factor responsible for stretching the polymer solution stream between the positive electrode and the negative electrode. If it is too low, the solution flowing through the nozzles will gather on the surface of the collector in the form of falling droplets of the solution. On the other hand, if it will be too high, the so-called supercritical tension will develop, resulting in several Taylor cones forming on the surface of the droplet, while their formation will lead to creation of varied fiber diameter, as well as too violent stretching, which along with insufficient transport of the solution may lead to pulse electrospinning. That is why, for each polymer solution, optimal tension value is always determined, at which the fiber reception process is stable. The mentioned phenomenon of the formation of the so-called Taylor cone is connected with tension. In the electrospinning process, after tension is applied, initially the charges of solution collecting on the surface are energetically offset by the surface tension of the solution. Along with the passage of time, density of charge on the droplet (solution) surface increases leading to the formation of the Taylor cone, of which the tip is always directed toward the second electrode. Under optimal conditions at the moment of exceeding critical polarization, meaning overcoming the forces of surface tension of the solution by the electrostatic field, the viscoelastic flow of the solution from the Taylor cone in the direction of the negative electrode begins. In effect, the charges on the cone surface begin to decrease. Then, the stream, at a specified distance, assumes a straight-line trajectory. As the straight-line stream moves toward the collector, at a specified distance depending on solution properties, a critical value develops and when it is exceeded a spiral movement of the solution stream begins, during which the strongest stretching of the stream occurs, connected with evaporation of the used solvent.

(iii) Solution properties

Combining a selected polymer material with the solvent enables the creation of the solution. The solution is characterized by a specified viscosity, conductivity pH, solvent evaporation speed, solution surface tension, etc. These properties are connected with one another, for example:

(a) Molecule mass of the polymer defines solution viscosity on which the fiber diameter depends.
(b) The obtained solution viscosity also depends on the density of the solvent used.
(c) Solution viscosity also depends on the concentration of polymer materials, the higher the concentration, the higher the solution viscosity, which may in turn be lowered by increasing solution temperature.
(d) The higher the solution temperature, the faster will the solvent evaporate.
(e) A change in solution temperature also leads to a change in solution conductivity.
(f) The higher the solution conductivity, the stronger will it be stretched by the electrostatic field.
(g) In most cases, as the percentage concentration of the polymer in a given solvent increases, the conductivity of the obtained solution decreases. However, there are polymer materials such as chitosan, for which along with an increase in concentration, solution conductivity also increases. Solution conductivity may in a significant way be altered by introducing organic additives, such as multiwalled carbon nanotubes, or inorganic additives such as silver nanoparticles.
(h) Since various solvents evaporate with different speed, often in order to obtain fibers with a porous structure or in order to obtain a higher stability of the process combinations of various solvents are used.

SOLUTION PROPERTIES AND THE PCL FIBER STRUCTURE—SELECTED EXPERIMENTAL EXAMPLES

In order to illustrate the influence of selected process conditions, solution properties as well as environmental conditions affecting the structure and properties of the obtained fibers, we provide some examples (Figs. 8.2.1–8.2.4), using: (i) polymers: polyvinyl alcohol (PVA) with Mw = 205,000, from Sigma Ulrich company, polycaprolactone (PCL) with Mw = 80,000 from Sigma Aldrich, (ii) reagents: DMF (C_3H_7NO), DMSO (C_2H_6OS), chloroform ($CHCl_3$), methanol (CH_3OH), distilled water, (H_2O), acetic acid (CH_3COOH), formic acid (CH_2O2) from Sigma Aldrich, (iii) additives: multiwalled carbon nanotubes

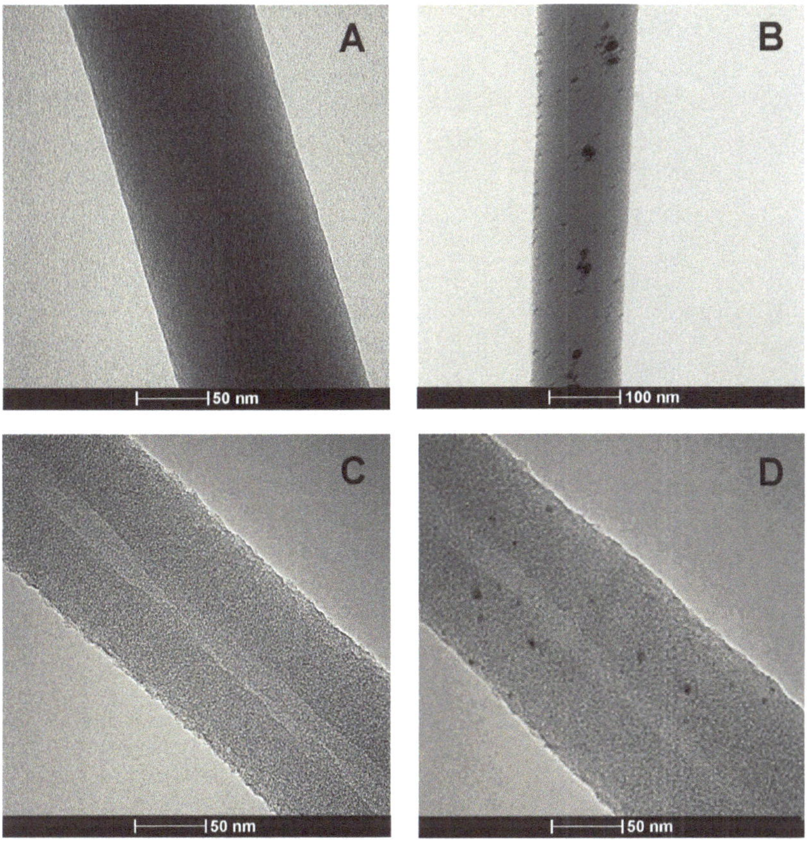

FIG. 8.2.1 Geometric characteristics of polycaprolactone fibers (PCL), either plain, or with (nano)particles added; (A) with a whole cross-section, obtained from a 10% solution of acetic acid and formic acid, with a mass ratio of 70:30; (B) containing silver (Ag) particles on the surface, pictures taken with the use of a transmission electron microscope TEM. (C) Geometric characteristics of core-shell fibers containing PCL on the shell as well as on the core; (D) with some silver (Ag) added on the shell, as well as PCL in the core. Pictures taken with the use of transmission electron microscope TEM.

(MWCNT) with a diameter of 50–100 mm and purity of 95.95%, titanium nanoparticles (Ti), with a diameter of 50–60 mm and purity of 95.95%, from Skyspring Nanomaterials, as well as silver nitrate ($AgNO_3$), with purity of 95% from Tantal. In order to prepare the solutions, equation indicated as follows was used to calculate the necessary amount of polymer materials with accuracy of 0.01 g, as well as solvents, which were calculated taking into account the viscosity of each individual solvent.

$$C_p = (m_s / m_r) \times 100\%$$

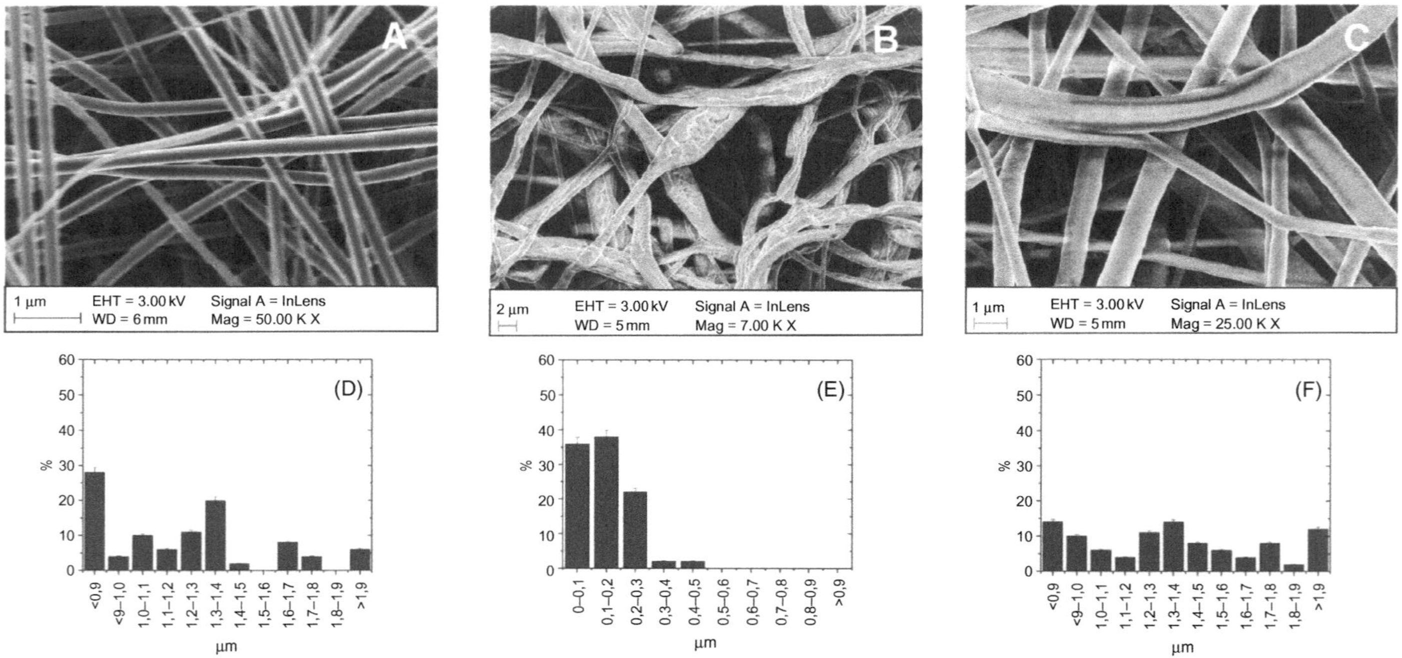

FIG. 8.2.2 Geometric characteristics of polycaprolactone fibers (PCL) obtained while using different solvents; (A) obtained from 10% solution of acetic acid and formic acid, with a mass ratio of 70:30; (B) obtained from 10% solution of chloroform and methanol, with a mass ratio of 70:30; (C) obtained from 10% solution of tetrahydrofuran and dimethyl sulfoxide, with a mass ratio of 70:30, pictures taken with the use of a scanning electron microscope SEM. Breakdown of the diameter of polycaprolactone fibers (PCL) obtained from (D) 10% solution of chloroform and methanol, with a mass ratio of 70:30, as shown in "A"; (E) from 10% solution of acetic acid and formic acid, with a mass ratio of 70:30, as shown in "B"; (F) from 10% solution of tetrahydrofuran and dimethyl sulfoxide, with a mass ratio of 70:30, as shown in "C."

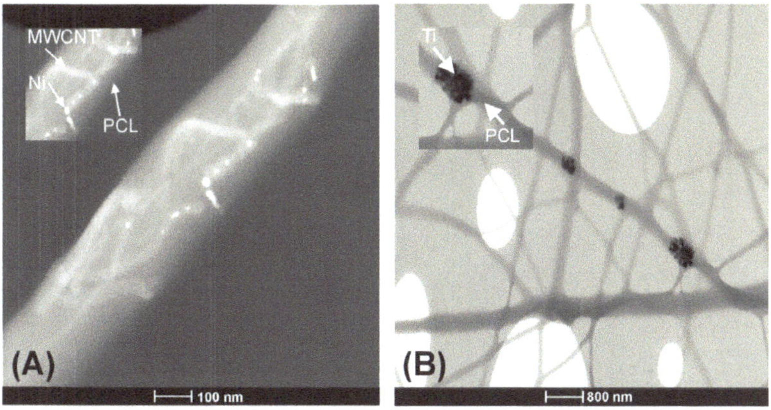

FIG. 8.2.3 Geometric characteristics of polycaprolactone fibers (PCL) obtained from 10% solution of acetic acid and formic acid, with a mass ratio of 70:30, with (A) multiwalled carbon nanotubes (MWCNT) added; (B) with titanium nanoparticles added. Pictures taken with the use of transmission electron microscope (TEM).

where C_p—% concentration expressed in "%," m_s—mass of the dissolved substance in g; m_r—mass of the solution in g. After dissolving the materials, the solutions were placed in containers for solutions in a device, with which flow and temperature were regulated, then they were subject to the impact of an electrostatic field transforming the solutions into fibers of a varied structure. The topography of the obtained nanofibers was analyzed by using scanning electron microscope (SEM) and transmission electron microscope (TEM).

The morphology of the nanofibers, and distribution of the nanoparticles varies between PCL, depending on the inclusion of the additives (Fig. 8.2.1). Fig. 8.2.2 illustrates how different solvents affect the morphology and properties of the obtained nanofibers. In Fig. 8.2.3, we show how the inclusion of either multiwalled carbon nanotubes (MWCNT), or titanium nanoparticles affects the electrospinning process and hence, nanofibers' morphology. Finally, Fig. 8.2.4 illustrates how the starting concentration of the material used for electrospin affects the morphology of the final product.

This chapter presented description of production of nanofibers, with the focus on the electrospinning method, underlines the versatility of the electrospinning process, and the examples cited here illustrate well, how changes in the composition of the starting material may affect the final product.

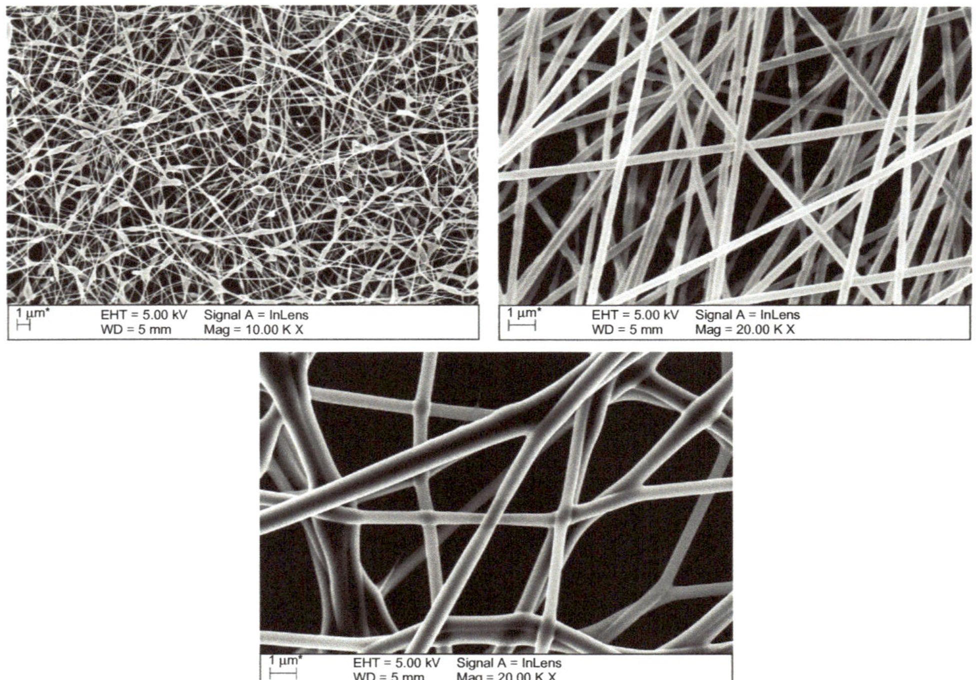

FIG. 8.2.4 Geometric characteristics of polyvinyl alcohol fibers (PVA) obtained from different concentrations of PVA's aqueous solutions; (A) from a 4% aqueous solution; (B) from a 7% aqueous solution; (C) from a 12% aqueous solution. Pictures taken with the use of scanning electron microscope (SEM).

REFERENCES

[1] Kwak S, Haider A, Gupta KC, Kim S, Kang IK. Micro/nano multilayered scaffolds of PLGA and collagen by alternately electrospinning for bone tissue engineering. Nanoscale Res Lett 2016;11(1):323. https://doi.org/10.1186/s11671-016-1532-4.

[2] Podgórski A, Bałazy A, Gradoń L. Application of nanofibers to improve the filtration efficiency of the most penetrating aerosol particles in fibrous filters. Chem Eng Sci 2006;61(20):6804–15. https://doi.org/10.1016/j.ces.2006.07.022.

[3] Khalil A, Kim JJ, Tuller HL, Rutledge GC, Hashaikeh R. Gas sensing behavior of electrospun nickel oxide nanofibers: Effect of morphology and microstructure. Sensors Actuators B Chem 2016;227:54–64. https://doi.org/10.1016/j.snb.2015.12.012.

[4] Chinnappan A, Baskar C, Baskar S, Ratheeshd G, Ramakrishna S. An overview of electrospun nanofibers and their application in energy storage, sensors and wearable/flexible electronics. J Mater Chem C 2017;5:12657–12673. https://doi.org/10.1039/c7tc03058d.

[5] Ilahi NA, Suryana R, Nurrosyid F, Linda Kusuma NT. Electrospinning titanium dioxide (TiO_2) nanofiber for dye sensitized solar cells based on Bryophyta as a sensitizer. J Phys Conf Ser 2017;795:012033 https://doi.org/10.1088/1742-6596/795/1/012033.

[6] Lin Z, Yao Y, Zhang X. Electrospun nanofibers for design and fabrication of electrocatalysts and electrolyte membranes for fuel cells. In: Electrospun Nanofibers for Energy and Environmental Applications. Berlin Heidelberg: Springer-Verlag; 2014. https://doi.org/10.1007/978-3-642-54160-52. (Chapter 2).

[7] Sun W, Cai Q, Li P, Deng X, Wei Y, Xu MM, Yang X. Post-draw PAN-PMMA nanofiber reinforced and toughened Bis-GMA dental restorative composite. Dent Mater 2010;26:873–80.

[8] Bajáková J, Chaloupek J, Lukáš D, Lacarin M. In: Drawing—the production of individual nanofibers by experimental method. Nanoconference 21–23.9.2011, Brno, Czech Republic, EU; 2011.

[9] Song Y, Wei J, Liu J. Template synthesis of hollow carbon nanofibers. Microsc Microanal 2015;21(Suppl. 3):https://doi.org/10.1017/S1431927615005747.

[10] Vasita R, Katti DS. Nanofibers and their applications in tissue engineering. Int J Nanomedicine 2006;1(1):15–30. PMC2426767 17722259.

[11] Liao H-S, Lin J, Liu Y, Huang P, Jin A, Chen X. Self-assembly mechanisms of nanofibers from peptide amphiphiles in solution and on substrate surfaces. Nanoscale 2016;8(31):14814–20. https://doi.org/10.1039/c6nr04672j.

[12] Augustine R, Kalarikkal N, Thomas S. Clogging-free electrospinning of polycaprolactone using acetic acid/acetone mixture. Polym-Plast Technol Eng 2016;55:518–29. https://doi.org/10.1080/03602559.2015.1036451.

[13] Kim GH, Cho Y-S, Kim WD. Stability analysis for multi-jets electrospinning process modified with a cylindrical electrode. Eur Polym J 2006;42:2031–8. https://doi.org/10.1016/j.eurpolymj.2006.01.026.

[14] Hudecki A, Gola J, Ghavami S, Skonieczna M, Markowski J, Likus W, Lewandowska M, Maziarz W, Los MJ. Structure and properties of slow-resorbing nanofibers obtained by (co-axial) electrospinning as tissue scaffolds in regenerative medicine. Peer J 2017;5:e4125. https://doi.org/10.7717/peerj.4125.eCollection2017.

[15] Wang C, Yan K-W, Lin Y-D, Hsieh PCH. Biodegradable core/shell fibers by coaxial electrospinning; processing fiber characterization and its application in sustained drug release. Macromolecules 2010;43:6389–97. https://doi.org/10.1021/ma100423x.

[16] Hong Z, ChenGuang Z, YunHui Z, GongWen T, Yan YX. Electrospinning of ultrafine core/shell fibers for biomedical applications. Chemistry 2010;53:1246–54.

[17] Khajavi R, Abbasipour M. Electrospinning as a versatile method for fabricating coreshell, hollow and porous nanofibers. Sci Iran F 2011;19(6):2029–34. https://doi.org/10.1016/j.scient.2012.10.037.

[18] Yang R, He J, Xu L, Yu J. Bubble-electrospinning for fabricating nanofibers. Polymer 2009;50(24):5846–50. https://doi.org/10.1016/j.polymer.2009.10.021.

Chapter 9

Application of Computed Tomography and Magnetic Resonance in 3D Modeling

Renata Wilk

Department of Anatomy, School of Health Sciences in Katowice, Medical University of Silesia, Katowice, Poland

ABBREVIATIONS

CT	computer tomography
FSE sequences	fast spin echo sequences
FLAIR	fluid attenuated inversion recovery
FID	free induction decay
HU	Hounsfield unit
IRS	inversion recovery sequences
MR	magnetic resonance
MRI	magnetic resonance imaging
MIP	maximum intensity projection
MPR	multiplanar reconstruction
NMR	nuclear magnetic resonance
PD	proton density
RF	radio frequency
SSD	shaded surface display
STIR	short TI inversion recovery
SSFSE	single shot fast spin echo
T	Tesla
TI	inversion time
TR	time of repetition
TE	time to echo
VE	virtual endoscopy

INTRODUCTION

Computed tomography is a diagnostic method that allows the use of X-rays to obtain body images in the form of cross-sections obtained in various planes. The obtained layers present the image of organs and tissues in relation to their

Stem Cells and Biomaterials for Regenerative Medicine. https://doi.org/10.1016/B978-0-12-812258-7.00009-5

spatial location and taking into account the exact internal structure. In the CT examination, the same phenomenon that was discovered and applied in classical X-ray, namely, the absorption of X-rays by tissues of different density is used. However, the sensitivity of the method and the use of specialized computer software cause that the tomographic sections provide much more detailed data on the structure of the human body in comparison to a conventional X-ray, and also allow us to perform 3D reconstruction of the structure we study. X-rays, a type of electromagnetic radiation, is generated by an X-ray tube, which consists of a vacuum glass tube and two electrodes with an opposite electrical charge. When using a high voltage on the cathode (with charge −), electrons are accelerated, which, reaching the tungsten anode (charge +), are braked on it, releasing electromagnetic radiation—X in this process. The maximum energy of this radiation depends on the energy of the initial electron. The voltage used in modern scanners ranges from 80 to 150 kVp [1].

CT SCANNER CONSTRUCTION

In contrast to the conventional X-ray examination, where a single X-ray tube is used as a source of radiation and a single detector, the CT scanner uses a mobile radiation source (X-ray tube) and a detector system (about 4000) placed around the circle called gantry. In fourth generation CT scanners, there are several moving X-ray tubes, while the detectors are placed on the entire circumference of the gantry in several rows. Gantra can be slanted toward the vertical axis by about 30 degrees to obtain oblique cross-sections and is used for examining, for example, the head and neck. Currently, the gantry inclination is abandoned because a spiral and multidetector tomography has emerged that allows obtaining similar images based on source data. The table with the patient moves during the examination by the gantry. The entire CT test set consists of a gantry; a movable table, which moves through a gantry, where the patient is placed; and a console allowing the operator to set the parameters of the examination and also to develop the obtained data after the test [2, 3].

During the test, the radiation source rotates around the patient inside the gantry, creating a narrow fan-shaped beam with different widths from (1–10 mm) depending on the used parallel collimators. The beam penetrates the patient's body, moving from different directions in the axial plane. This allows to irradiate exactly this part of the patient's body, which cross-section we want to obtain at a given moment. Detectors placed on the opposite side to the radiation source record the image from a given cross-section in the form of one shot in one setting of the X-ray tube.

OBTAINING CT IMAGE

The principle of obtaining images in CT is based on different degrees of radiation absorption by individual tissues depending on their density and slide thickness.

The lowest degree of radiation absorption shows air, so tissues filled with air such as lungs are dark on CT images. Bones that show high density and at the same time a high rate of radiation absorption are bright. The remaining tissues are presented on CT images in various shades of gray (darker or brighter) [2–4].

After collecting and recording signals by the detectors of the scanner, obtained data should be preprocessed. First of all in the range of compensation of detectors inhomogeneities and also correction of the radiation beam hardening effects. The data obtained after the initial processing are called the raw data (CT raw data), from which image data files are obtained in the process of image reconstruction. The reconstruction, which takes place during the measurement, is a preliminary reconstruction, whereas the secondary reconstruction leads to the creation of a tomographic image [5].

For the reconstruction of the image in the form of 2D raster images, a mathematical transformation is used, referred to as the inverse transform of Radon [3]. However, there are other methods of image reconstruction, selection of the appropriate method allows to avoid the appearance of artifacts. The initial step of reconstruction is choosing the matrix of detectors on which it is created. Currently, matrices with the size of 512×512 pixels are most commonly used, but they can be different (256×256 or 1024×1024). In this case, the pixel is a unit of a two-dimensional image; however, given that the layer used in the CT scan has a specific thickness, the equivalent of a pixel in space is a voxel. The size of the cube depends on the selected imaging field (FOV), matrix and layer thickness, usually within the range of 0.6–1 mm. The shape of the voxel is not cubic but has the shape of a rectangular prism, and the size of the pixel on the cross-section is 10–20 times smaller than the thickness of the layer. The next step in image reconstruction is to determine the radiation attenuation for each pixel, with averaging resulting from the shape of the voxel. The averages mainly concern the differences in the degree of radiation absorption by different tissues occurring on the entire thickness of the layer. In order to improve the image quality, also mathematical algorithms are used, the task of which is min. Improvement of edge contours and noise reduction. Soft and hard algorithms are used the most often. The low-resolution soft algorithm is used for imaging organs with a smaller difference in the degree of radiation absorption and thus exhibiting weaker contrast in the image. Objects with a significant difference in the degree of radiation absorption require the use of a hard, high-resolution algorithm that allows the edge to be strengthened while not allowing the distinction between weakly contrasted structures [3]

The quality of the CT images obtained depends on the different types of resolutions used in creating images. Contrast resolution determines the degree of discrimination of tissues with a small difference in density. Spatial resolution allows to recognize small objects in the image, which in this case depends on the size of the voxel. To improve it, it is necessary to reduce the layer thickness, increase the number of cross-sections used, and increase the radiation dose. The time resolution, however, depends on the time necessary to collect all the necessary data. The shorter the test time, the lower the probability of the so-called motion artifacts resulting from uncontrolled patient movements [4].

CT image reconstruction is based on a linear coefficient of radiation absorption and the Hounsfield scale. The linear radiation absorption coefficient determines the part of the radiation beam, which has been absorbed or dispersed when passing through the examined tissue, per unit of its thickness. In general, factor is proportional to the density of the tissue. Each given is assigned the appropriate numerical value, the CT number corresponding to the degree of radiation attenuation passing through its area. To determine the number of CT, the appropriate formula is used:

$$\text{CT number} = 1000 \times (\mu - \mu_{\text{water}}) / \mu_{\text{water}}$$

where μ is the coefficient of radiation absorption for a given tissue, μ_{water} is the attenuation factor for water [2] to make it simple, the CT number has been called the Hounsfield unit (HU). The HU is related to the linear absorption coefficient and is equal to 0.1% of the difference between the absorption coefficient for a given tissue and the linear absorption coefficient for water. Gases, including air, represent the lowest value "−1000" HU, fat "−50" HU, water "0" HU; Gases, including air, assume the lowest value of "−1000" HU; fat "−50" HU; water "0" HU; soft tissues from "+30" to "+80" HU; and bone "+1500" HU. Metal implants "+3000" HU, however, the upper limit of the scale has not been established. The range of CT numbers depends on the coding system used in the CT scanner and the number of bits in a pixel, a typical 12-bit system allows for coding in the range from −1024 to +3071 HU (Fig. 9.1).

The use of the full range of gray scale HUs, that is, about 4000 HU would be nondiagnostic, because the human eye cannot distinguish all the shades of gray presented, and thus the distinction of structures with small differences in HUs is no longer possible.

About 150 shades of gray is a range that can be distinguished for a trained eye. Therefore, imaging studies use so-called windows that allow for presenting the same images in different ranges of gray scale according to HUs. The window has two parameters that determine it: width—conditioning the contrast, determines the number of HU shown on the screen, and level—otherwise the center of the image defining the brightness by Hounsfield number in the center

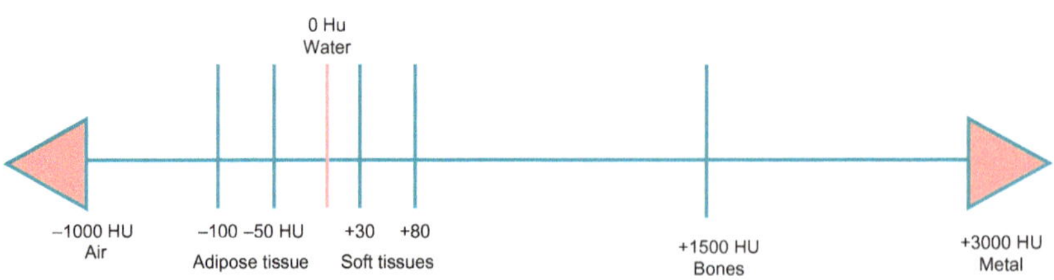

FIG. 9.1 Distribution of Hounsfield units with reference to the base unit—water absorption. *(according to Kelly B, Bickle IC, Diagnostyka obrazowa. Walecki J. (ed.), Urban & Partner, Wrocław 2008.)*

of the window width. Window settings are defined for individual tissues and can be determined using specific ranges according to HUs. To calculate the upper and lower limits of the window range, the appropriate formulas are used:

Upper limit of window range = level in HU + width/2;
Lower limit of window range = level in HU − width/2,

for example, a lung window with a level of −650 and a width of 1500 will have ranges from 100 to −1400 HU [2, 3].

The most commonly used windows in computed tomography of chest are: soft tissue window in the range from 240 to −160 HU, lung window from 100 to −1400 HU, and bone window from 1400 to −600 HU. The brain examination uses a window with a range of 85 to −15 HU (Fig. 9.2). Diagnostic windows are more dependent on the tissue being examined. Window settings are given as parameters for specific tests. The upper limit of the range means the maximum brightness of the pixel, while the lower limit of the pixel is the smallest brightness (darkness). The values of window settings for specific organs and tissues are given in table (Table 9.1).

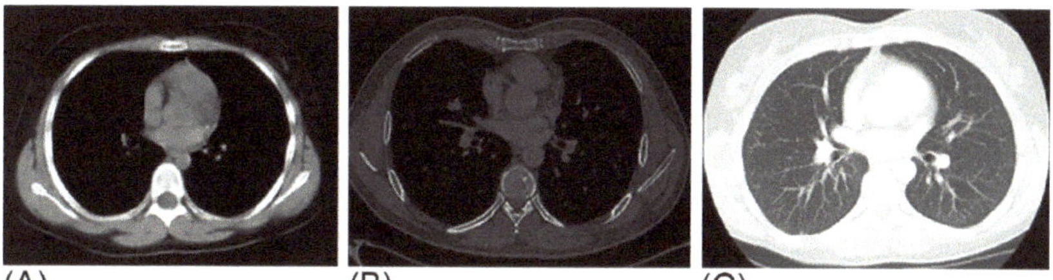

FIG. 9.2 Windows used in chest computed tomography. (A) Soft tissue window, (B) Bone window, and (C) Lung window. *(CT scans, courtesy of the Radiology Department of The Independent Public Clinical Hospital No. 6 Medical University of Silesia in Katowice. John Paul II Upper Silesian Child Health Centre.)*

TABLE 9.1 The Window Settings Applied in Examination of Various Tissues

Tissue/Organ	Window Width	Window Level
Lungs	1200–1500	−600 to −650
Brain	80–100	35
Abdominal organs	300–350	40
Bones	2000–2400	500
Vertebral column	350–400	35–40
Mediastinum organs	400	40

SPIRAL (HELICAL) COMPUTER TOMOGRAPHY

It is based on continuous scanning with an X-ray tube that rotates around the patient's body while the table is being moved by the gantry. As a result, the data are collected continuously from every point of the patient's body. The examination takes a relatively long time, so considering time of one rotation of the X-ray tube is even over 100 s, it is necessary to use a X-ray tube with a large heat capacity and adequate cooling. Axial cross-sections can be obtained from any level with any space, because data acquisition takes place by volume. The distance between the layers is defined as the distance of reconstruction or increment. At the collimation of the section, the thickness of the layer and the table travel are selected independent of each other. Pitch is the jump factor, which determines the speed of the table travel, per one turn of the X-ray tube in relation to the collimation of the layer. It should be kept in mind that the reconstruction interval does not correspond to the collimation of the layer; it only means the location of the reconstructed sections obtained from the volume data in relation to each other [2]. The pitches numbers the most commonly used in imaging examinations are from 1.5 to 2; however, smaller ones can be used as well. It is important to remember that using the pitch values below 1 causes overlapping of the layers and increases the radiation dose received by the patient [3]. Values above 2 may lead to inadequate depiction of the examined object and the artifacts appearance [2–4]. The choice of table movement in clinical practice depends on the scope of the examination and the time necessary for its implementation. The maximum available time should not exceed 30 s, because the examination requires breath hold. Hence, the applied pitch value is often greater than 1, thanks to which the radiation dose is reduced and the highest spatial resolution at the same time is achieved. Image reconstruction in spiral tomography is carried out using the same algorithm that is used in classical tomography. The reconstruction spacing should be chosen so that 30% of the cross-sections overlap, especially when imaging small objects. In order to obtain spatial reconstructions, it would be recommended that the sections overlap up to 50% to avoid omitting those changes that locate on the border of adjacent layers. Presentation of CT images from spiral tomography is done in a similar way as conventional tomography; however, the number of images obtained in the study, at the same time, is many times greater. Most often, not all of the images are recorded, but only every 2, 3, or 4, other images, especially in the area of pathological changes, are retained for possible subsequent evaluation [2, 3].

MULTIDETECTOR COMPUTER TOMOGRAPHY

It is also referred to as multilayered or volume tomography. It offers an imaging system that provides a greater range of access to the examined structures, shortens the time of examination as well as reduces the collimation of the layer. The structure of a multirow CT scanner is based on 3rd generation scanners, with a rotating radiation source and detectors placed around the patient's body.

However, in place of one row of detectors, initially four of them were used, enabling simultaneous acquisition of data from multiple layers. Currently, CT scanners with 64 or 128 rows of detectors are used.

There are several types of detectors used in multidetector tomography: The best known are the Uniform or Matrix Detectors, in which the rows of detectors are of equal width, the layer can be adjusted by collimating the beam and activating the appropriate number of layers of detectors, the width of one detector row is 1.25 mm. On a matrix equipped with four rows of detectors depending on the collimation of the beam and activation of the detectors, the test is performed with a 4×1.25 mm collimation. When activating successive rows of detectors, you can get 4×2.5, 4×3.75 mm or more.

Nonuniform or Variable Detectors are made up of rows of detectors with different widths, wider at the circumference and tapering inwards. Such a method is justified by the shape of the radiation beam, in which detectors located on the perimeter absorb radiation falling diagonally. Perpendicular partitions in this section would block part of the radiation beam and reduce its detection efficiency. Obtained collimations are from 2×0.5 mm through 4×1 mm to 4×2.5 mm.

Hybrid Detectors are made of rows of detectors of different widths; however, only two dimensions are present here, wider at the circumference and narrower in the middle part. The entire table has a dimension of 32 mm, the width of each row of detectors is 1 mm, while the 2 mm midsection has been divided into four rows with a width of 0.5 mm. Collimation ranges from 4×0.5 to 4×8 mm.

Multidetector tomography, thanks to its structure, allows for simultaneous imaging of the whole selected part of the body (abdomen or chest) or body elements passing from one part of the body to the other like the spine or aorta. The method allows for shortening the time of examination while maintaining the appropriate diagnostic value. The thickness of the layer can be reduced while providing identical resolution in all cross-sections. Examinations allow also for the reconstruction of the cross-section obtained from any point of the patient's body. Currently, the concept of multilayer tomography is increasingly used. It refers to the number of cross-sectional layers obtained per unit time and does not necessarily refer to the number of rows of detectors in the CT scanner, because additional layers can be obtained by interpolating the data obtained during the incomplete rotation of the lamp [2, 3].

CT CONTRAST AGENTS

Despite the far-advanced modernization of CT scanners and introduction of new solutions facilitating the conduct of the study itself, for diagnostic purposes it is still necessary to precisely distinguish tissues with a similar degree of radiation absorption. For this purpose, contrast agents are used to increase the difference between adjacent tissues and also provide information on the biochemical characteristics of the tissue under study and to assess its function to some extent.

Contrast media owe their properties to the content of chemical elements with a high atomic number with simultaneously low harmfulness to the body and the ease of their removal in a few hours, usually through the kidneys. The contrast is usually administered intravenously using an automatic syringe with a programmable amount and adjustable flow in order to spread it evenly during the examination. However, other routes of administration are possible: oral, intra-arterial, or rectal. Current contrast agents are mainly based on iodine (atomic number 53), especially in compounds using iodine covalent bonds. Contrasting iodine agents can be divided into ionic and nonionic ones. Ionic iodine contrast agents are highly harmful and because they are characterized by high osmolarity, they can lead to kidney damage and other disorders such as vasodilatation, bradycardia, and increased pressure in the pulmonary vessels. Nonionic iodine contrast agents with low osmolarity have a better tolerance; they do not cause such serious complications as the previous ones. Among the nonionic contrast agents, there are also those whose osmolarity is close to blood but with a higher viscosity. Viscosity is crucial here both for the correct distribution of contrast during its administration and for the rate of its removal from the body after the examination. The best-tolerated contrast agents are those with low osmolarity, while the low viscosity ensures the right rate of agent delivery during the test. In addition, the high concentration of iodine in the administered agent allows for a reduction in the volume necessary for administration. Due to the possible complications associated with the occurrence of anaphylactic reactions of varying severity, the administration of contrast media requires proper preparation and is not always possible. Contraindication to the administration of contrast is, among others, renal failure, diabetes, acute myocardial infarction, kidney transplantation [1, 2] (Fig. 9.3).

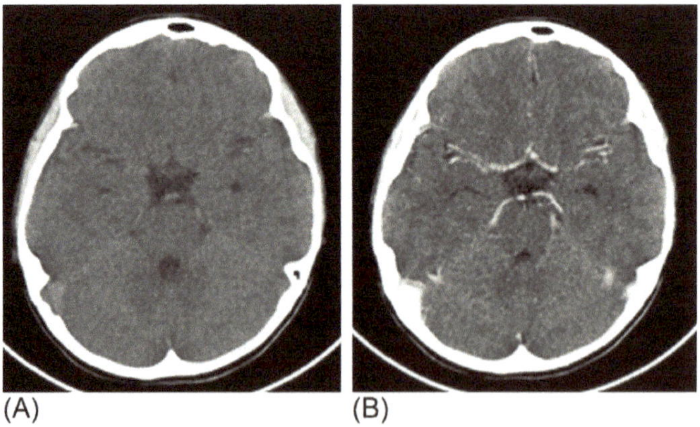

FIG. 9.3 Head CT images (A) without contrast, (B) with contrast. *(CT scans, courtesy of the Radiology Department of The Independent Public Clinical Hospital No. 6 Medical University of Silesia in Katowice. John Paul II Upper Silesian Child Health Centre.)*

PRESENTATION OF IMAGES FROM THE CT STUDY AS TWO DIMENSIONAL

Cross-sections are the basic sections used to represent images based on a CT scan.

The most common format generated by CT scanners is DIACOM that allows for viewing images, compare images, and make corrections. It allows to choose the appropriate diagnostic window based on raw data and also to perform some measurements important in diagnostics.

3D IMAGING IN COMPUTED TOMOGRAPHY

Obtaining images in planes other than lateral ones requires changing the arrangement of the analyzed voxels, which fill a specific volume of the part of the body being scanned, using appropriate algorithms representing particular sections. The method of multiplanar reconstruction—(MPR) is used for this purpose. Currently, the 3D study uses images coming from spiral or multidetector tomography, which provide volumetric data, allowing for reconstruction in various planes, and as a result, 3D images obtained on the basis of the same raw data (Fig. 9.4).

Surface imaging is possible using the SSD (shaded surface display), which is an algorithm that adapts to the surface of anatomical structures with the value

FIG. 9.4 CT 3D reconstruction of thorax. *(CT scans, courtesy of the Radiology Department of The Independent Public Clinical Hospital No. 6 Medical University of Silesia in Katowice. John Paul II Upper Silesian Child Health Centre.)*

of the attenuation factor greater than that chosen by the operator. After selecting the appropriate ranges, the choice of the object can be done automatically, semi-automatically, or manually. The procedure is important in preoperative planning of vascular system or orthopedic surgeries. Its disadvantage is the imaging of only the surface of the chosen structure without information about its interior.

Virtual endoscopy (VE) provides data for simulation of images that can be obtained during a classic endoscopic examination and is usually based on differences in the coefficients of radiation absorption between the gas filling the intestine or other tubular organ and its wall.

Maximum intensity projection (MIP) or minimum intensity projection are techniques that allow for displaying only these pixels that exhibit very high or very low levels of radiation attenuation compared to the surrounding ones. These methods are used in imaging of the vascular system, in which the contrast agent filling in the vessels has high radiation absorption coefficient but also for imaging objects, exhibiting poor absorption coefficient, for example, gallbladder or not contrasted, widened urinary tracts [2, 6].

The methods given here allow obtaining 3D reconstruction of images, important especially in the diagnosis of internal organs in order to eliminate the need for more invasive methods. For example, in studies necessary for planning liver transplants, 3D reconstructions mainly concern vascular system in the arterial and venous phase, mainly depicting the hepatic artery and portal vein. In addition, images of the liver parenchyma itself for possible damage as well as for the accurate mapping of the vascular segments necessary for the transplant.

In preoperative imaging of the pancreas and bile ducts in cases of pancreatic cancer diagnosis, the study allows to determine the topography of pancreatic blood vessels in relation to the tumor as well as the tumor and organs adjacent to it such as the duodenum, liver, or transverse colon. It is also important to assess the degree of bile ducts as well as pancreatic duct dilatation.

CT urography is another examination allowing for replacement of classical urography. The use of the urinary tract 3D reconstruction allows easier localization of the place of obstruction and possible anatomical variations of the urinary tract or vascular system. In addition, pathological changes including the bladder wall and inside the kidney are visualized.

3D view of the gastrointestinal tract allows for accurate visualization of pathological changes as polyps or diverticulitis but also ischemic or inflammatory changes [6, 7].

MAGNETIC RESONANCE

MR is a different diagnostic method than classical radiography, which allows for obtaining images of the human body. For obtaining data, a strong magnetic field is used in which the patient is placed. Then a strong radiofrequency impulse is sent and absorbed by the patient's body. After the radioimpulse is interrupted, the energy absorbed by the patient's body is transferred in the form of a radio wave, which can be used for image reconstruction.

Construction of a Nuclear Magnetic Resonance (NMR) Scanner

The NMR Scanner consists of a strong magnet. The types of magnets used in medicine have magnetic field induction from 0.3 to 3T (1T (tesla) = 10.000 Gs (Gauss). The strength of the magnetic field depends on the type of magnet used.

There are permanent magnets with a low magnetic field strength value of 0.2–0.6T, which can be used in open cameras. They exhibit magnetization that persists constantly, so they require less energy, but their disadvantage is a very high mass reaching 100 tons, which requires specially adapted rooms. The device is made of open yokes and a table on which the patient is placed. The magnetic field is generated from above and below the patient. The advantage of these magnets is the ability to perform tests in patients with claustrophobia and obesity because they are more spacious than classic ones [2, 7].

The second group of magnets are electromagnets, which can be further divided into resistive and superconducting magnets.

The most commonly used magnets are superconducting magnets. In these magnets, an electric conductor cooled to the superconducting temperature (4K, 4 Kelvins) is used, so that it does not show any resistance to the electric current, which allows it to generate a constant magnetic field. They have the shape of a cylinder, around which there is magnet and a system of transmitting and receiving coils. They require very effective cooling because they get extremely hot. Helium is most often used for cooling, and superconductors are kept at a temperature of liquid helium. An increase in the temperature of the superconductor causes the loss of its properties, there is resistance in the flow of current and further increase in temperature, the coolant goes into gas and can lead to an explosion if it does not find its way out. Electromagnets used in medical imaging examinations produce a magnetic field with an induction of 1.5–3T. Their advantage is the generation of a strong magnetic field with very high uniformity. During work with strong magnets, it is important to maintain certain safety rules. First of all not to bring metal objects to the MRI examination room because they can be pulled into the gantry. Electronic devices may affect the homogeneity of the magnetic field and cause artifacts during the study.

In addition to the main magnet, so-called coils are used. They are the overlays on individual parts of the patient's body, which are supposed to improve the process of reading data (receiving coils) or transmit radio frequency (RF) pulses and then receive signals from a given part of the body after the end of its transmission.

Coils can be divided into:

- *Volume coils*—completely surrounding a given part of the body, they usually act as transmitting and receiving coils;

- *Shimming coils*—which task is to improve the homogeneity of the magnetic field in a given area;
- *Gradient coils*—Gradients (gradient coils) allow to create an additional magnetic field with a specific gradient facilitating the acquisition of data from a specific place—this applies both to the selection of the layer and the point within it;
- *Surface coils* adhere directly to the test area, they are usually receiving coils, they allow the study of the surrounding areas in their immediate vicinity [2].

Obtaining a Signal in MR

As in the case of CT, the image in the MRI is collected from small fragments of space (voxels), from which data on the signal strength characteristic for a given tissue and the selected sequence are collected. The MR device is designed to measure the resonance signal from each of these space fragments, locate them in 3D space, and place them on a 125×125 or 512×512 in. matrix to get a visible image. The basis for receiving the signal in MR is to place the patient in the magnetic field. The magnetic field then acts on the atoms that build the organism and specifically on their atomic nuclei. In every part of the body space we are exploring, or voxels, there are certain atoms of elements that can be used to obtain the signal that is the basis of NMR. The nucleus of the atom consists of nucleons, or protons and neutrons, but only the protons are subject to the action of the magnetic field, because they have a positive electric charge and only in those nuclei in which the number of nucleons is odd. An atom that fulfills this condition and is very abundant in the body is hydrogen, which has only one proton in its nucleus, so to facilitate its behavior is described on the example of a proton. Each proton rotates around its own axis, so it has a spin and due to the electric charge moving in this motion, it generates electric current and thus also the magnetic field. After placing the protons in the external magnetic field, they will be able to set in it in two directions: in the direction of the magnetic field—some of the protons align with the direction the magnetic field (parallel) and some align opposite to direction of the field (antiparallel). This setting depends on the energy level of individual protons. If the proton energy level is higher, they are set antiparallel to the field direction, if lower—parallel to the direction of the field. Generally, the opposite to the field direction requires higher energy than the direction with the field. The difference between protons in direction with the field and opposite to the field is 10,000,007–10,000,000, so the parallel position prevails. In addition, the movement of the proton around its own axis takes place in the form of a precession motion, outlining the shape of the cone around the axis of rotation of the proton. This movement takes place at a high frequency, which depends on the intensity of the magnetic field. The frequency of precession movement is determined by Larmor's equation:

$$\omega_0 = \gamma B_0$$

where B_0 means magnetic induction (in tesla (T)), γ is a gyromagnetic coefficient, and ω_0 is the frequency of precession.

The frequency depends on the magnetic field induction but also on the specific gyromagnetic coefficient of the substance. For hydrogen, the gyromagnetic coefficient γ is 42.59 MHz/T.

The magnetic moment of the protons turned with the magnetic field direction is neutralized by the magnetic moment of the protons facing the opposite to the field, but because the number of protons turned with the magnetic field is more, the resultant magnetic moment (total) is directed with the working magnetic field. The magnetization obtained in this way is referred to as longitudinal magnetization consistent with the direction of the active magnetic field. The use of RF pulse disrupts the precession of protons by providing an additional dose of energy. However, for the energy to be taken over by the protons, the RF pulse must have the same frequency as the precession frequency of the proton, which can be calculated according to the Larmor equation. The transmission of energy takes place on the principle of resonance of waves with the same frequencies. Protons that have taken over the energy change their position from parallel to antiparallel, but also their precession frequency undergoes synchronization (they precess in a compatible phase), finally adopting a direction other than longitudinal magnetization consistent with the direction of the magnetic field. We obtain transverse magnetization, the vector of which is directed to the side, it also performs a precession movement about the axis consistent with the direction of the magnetic field and the precession frequency of the hydrogen protons. A change in the magnetic field is the source of energy to be picked up as a MR signal. A very important issue here is how to locate the signal within the body part being examined. When using a uniform magnetic field in the body part to be tested, it will not be possible, but the use of different magnetic field values for different points of the body allows it. Data derived from the body in the form of an MR signal can be related to the distribution of the applied magnetic field of varying intensity, knowing that the frequency of precession is correlated with it. Frequency values of the MR signal allow precise location of individual points in the patient's body and image reconstruction [2, 7, 8].

T1 and T2 Relaxation

After switching off the RF signal, the protons that have absorbed energy start to return to the equilibrium, this process is called relaxation. At first the transverse magnetization begins to disappear, so we have to deal with transverse relaxation and in the second place the longitudinal magnetization is reproduced, so the phenomenon of longitudinal relaxation appears. Transverse relaxation involves the loss of synchronization of individual protons movements, which is referred to as protons exits from the phase. The pace of this process depends on several factors, including the influences of individual protons on each other, because they are the source of small magnetic fields and the external magnetic field has certain heterogeneity. Hence, these influences are characteristic for individual tissues. The T2 relaxation time is a process that time cannot be precisely defined as one value. It is reported that the decrease in transverse magnetization to 37% of its initial value is defined as the T2 relaxation time. This time depends mainly on interspinal interactions and therefore is also referred to as spin-spin relaxation. Time T2 in the living tissue determines the range from 30 to 150 ms. Time T2 depends on the heterogeneity of the magnetic field, with the high heterogeneity of the external magnetic field and the heterogeneity of local magnetic fields, the time T2 is shorter. Longitudinal relaxation involves the return of protons to a lower energy level after switching off the RF signal, which results in a return from antiparallel to parallel. In this way longitudinal magnetization is reproduced, which is why this phenomenon is called longitudinal relaxation—T1. T1 specifies the time constant necessary to reproduce the longitudinal magnetization up to 63% of its initial value. The energy that the protons give off during this process goes to the lattice of surrounding atoms, hence the determination of T1 relaxation as a spin-lattice relaxation. Time T1 is from 300 to 2000 ms, so it is even 10 times longer than time T2. T1 is strongly dependent on the type of tissue and the value of the magnetic field, getting longer in a stronger magnetic field. Individual times vary depending on the tissue they come from, but this does not give you the opportunity to determine what tissue you are dealing with. It is reported that water and other liquids have long relaxation times for both T1 and T2 while adipose tissue has short [2, 7, 8].

K Space and Image Creation in MR

Raw data from the MR signal are collected in the so-called *K* space being an image of a 2-dimensional matrix filled with data. These data are subjected to mathematical transformation using an inverted Fourier transform to obtain the final image. A single cross-section in MR is created from data collected in a defined plane of *k* space in real time. Each point of this space encodes three values; frequency and phase as *x* and *y* coordinates and signal intensity, which corresponds to brightness. The central part of the *K* space encodes the information about contrast, while the part of the space on the perimeter, the spatial resolution of the image [8].

Image Acquisition in MR

It can be either 2-dimensional or 3-dimensional. There are three types of magnetic field gradients emitted by gradient coils that allow for differentiation of individual fragments of the examined cross-section. The gradient of the *Gz* (along *z*-axis) layer selection used during the excitation is intended to excite the proton resonance in only one layer. The *Gx* (along *x*-axis) frequency coding gradient allows the frequency of precession of nuclei in a given layer to be varied, sent during reading along the *x* axis. The gradient encoding the phase of *Gy* (along *y*-axis), aiming at the differentiation of the phase in which protons are precessing in a given layer, is activated just before the reading, while during the reading the same gradient will be used to correctly read the data. Using the read gradients in 2D acquisition, the information provided as an echo includes the source position relative to the *x*-axis encoded in the frequency of the signal and the *y*-axis in its phase. With 3D acquisition, a fixed gradient *Gz* is replaced by a variable gradient that allows you to excite a specific volume of the object being examined. The information comes from all three axes [8].

BASIC SEQUENCES USED IN MR STUDIES

Spin Echo Sequences

This is the basic sequence used in MR. It is based on sending two successive RF pulses, the first is defined as the 90 degrees pulse, the second 180 degrees pulse. The 90 degrees pulse is sent to flip the direction of the longitudinal magnetization vector by 90 degrees, as a result of which we get a transverse magnetization vector and a reduced longitudinal magnetization vector. A magnetization vector is created with an intermediate direction in relation to the two basic ones. After switching off the RF pulse, the protons leave the phase and the magnetization vector returns to the initial state, that is, longitudinal magnetization is reproduced. The received signal during this process is referred to as free induction decay (FID). It results from the dispersion of the proton precession phase. This signal gradually disappears over time, but its frequency does not change and is consistent with the precession frequency and its intensity depends on the density of protons in a given tissue.

The time from applying the pulse from the reading of the signal is time to echo (TE). The short TE is defined as less than 30ms and the long time is above 80ms. The time between individual 90 degrees pulses is referred to as the time of repetition (TR). This time, if it is long enough, allows to reproduce the longitudinal magnetization and depending on the T1 relaxation time of the given tissue, this signal will change. The TR is considered to be short if it is less than 500ms. The long time TR is measured above 1500ms. By shortening TR, it is possible to differentiate the signal coming from tissues with a different relaxation time T1; therefore, the image obtained in this way is called T1-weighted image.

The use of a very long TR means that the longitudinal relaxation is no longer relevant, but there are differences visible in the images resulting only from the proton density (PD) of the tissue in which the signal was measured. Thus, the image is determined to be dependent on the PD—PD weighted image.

Dependent T2 images are obtained by applying an additional RF 180 degrees pulse sent exactly in the middle of time between the 90 degrees pulse and the TE time. This time, for simplicity, is referred to as TE/2. This impulse is designed to reverse the direction of protons precession by 180 degrees and also synchronize their movement. Because the first 90 degrees impulse caused the occurrence of transverse magnetization, the 180 degrees impulse will cause its additional reinforcement. Therefore the T2 relaxation lengthens after applying the impulse 180 degrees, while removing the heterogeneity of the external magnetic field, but the intensity of the signal decreases with each subsequent spin echo reading [8].

Typical Images Used in the MR

Depending on the TR and TE used, the intensity of the signal will depend on various factors. With the use of short TR and short TE, the intensity of the signal will depend on the pace of longitudinal magnetization, that is, it will be T1-weighted image. The shorter the T1 relaxation is of a given tissue, the more intense the signal will come from it. The times used for T1 weighted images are: TR from 400 to 500 ms and TE from 15 to 25 ms.

With the use of long TR and long TE, differences in T1 will not be visible, because all longitudinal magnetization has been reconstructed so the only thing that influences the differences in the study is related to the different transverse relaxation time for individual tissues. The times used for T2-weighted images are TR—up to 2000 ms and TE—from 100 to 150 ms.

With the use of long TR and short TE, longitudinal magnetization will be completely reconstructed, so the relaxation time T1 will not affect the differences in the MR image. A short TE will, however, make the differences related to the transverse T2 relaxation not visible. Thus, the image is independent of T1 and independent of T2. The factor that in this case affects changes is the PD in the examined tissue. The more protons there are in the tissue, the stronger the signal.

In the MR images, from the practical point of view, the following dependences can be noticed. The liquid environment, for example, showing the cerebrospinal fluid on T1 images, will appear in black, the T2 and PD images will be bright [2, 8] (Figs. 9.5–9.7).

Fast Spin Echo Sequences

They are used primarily to shorten the time necessary to perform the examination. It involves the use of standard pulse of 90 degrees and then a series of pulses of 180 degrees. As a result, we get several echoes at the same TR. The number of echoes obtained during TR is referred to as echo train length or turbo factor; it usually ranges from 4 to 32 in a standard examination but can increase up to 200. At 200 pulses or more, this sequence is referred to as single shot fast spin echo (SSFSE) and they allow the imaging of organs in motion. The FSE sequences, however, require the use of strong gradient coils as well as a uniform magnetic field. Coils are necessary to change the precession phase of individual protons. They allow the simultaneous scanning of multiple lines within the space k, which extends the TR and improves the signal-to-noise ratio by strengthening the longitudinal magnetization. Spatial resolution, in this case, can be improved by making changes in the gradient encoding phase [9].

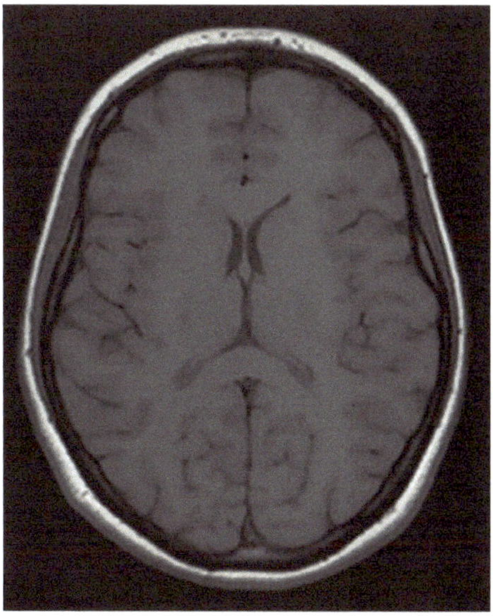

FIG. 9.5 Brain image weighted T1: the cerebrospinal fluid in cerebral ventricles is *black*, the *gray* matter is *dark*, the *white* matter is brighter than the *gray* matter. *(MRI scans, courtesy of the Mediskan, Zawiszy Czarnego 7A, 40–872 Katowice.)*

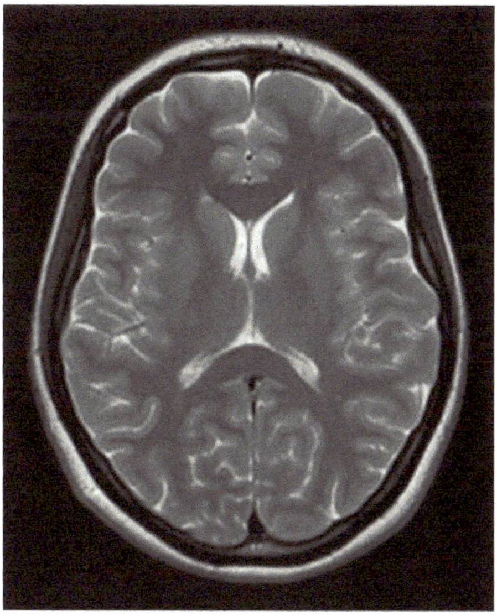

FIG. 9.6 Brain image weighted T2: the cerebrospinal fluid is bright, the *gray* matter is brighter than the *white* matter. *(MRI scans, courtesy of the Mediskan, Zawiszy Czarnego 7A, 40–872 Katowice.)*

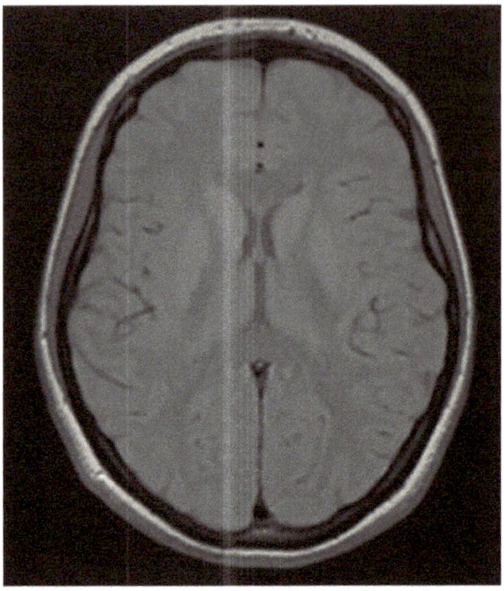

FIG. 9.7 Brain image weighted PD: cerebrospinal fluid is bright, *gray* matter is brighter than *white* matter due to its higher water content. *(MRI scans, courtesy of the Mediskan, Zawiszy Czarnego 7A, 40–872 Katowice.)*

Inversion Recovery Sequences

They consist of using 180 degrees pulse before the standard 90 degree pulse for the spin echo sequence. The time before the 90 degrees impulse is referred to as the inversion time (TI). The inversion impulse is aimed at reversing the direction of longitudinal magnetization in the direction opposite to the direction of the active magnetic field. At TI, the tissue undergoes relaxation T1. After application of the 90 degrees pulse, divergent signals appear from tissues that differ in the T1 relaxation time. The better contrast between tissues can be obtained by the appropriate selection of the TI parameter but also TR and TE [10].

Short TI Inversion Recovery

One of the examples of such a sequence is Short TI Inversion Recovery (STIR), in which using the appropriately short TI it is possible to suppress the signal from tissues with short T1, in this case fat. The time used here for a 1.5-T magnet is 160–180 ms. The method can be used for both T1- and T2-weighted images on which T2 is particularly useful for detecting edema changes. The signal from fat and water in T2 is very strong, so removing the signal from the fat causes the signal from the water to become clearer [11].

Fluid Attenuated Inversion Recovery (FLAIR)

FLAIR is another method using inversion recovery. The method allows for (using a long TI time on T2 images) obtaining an attenuation of signal from the

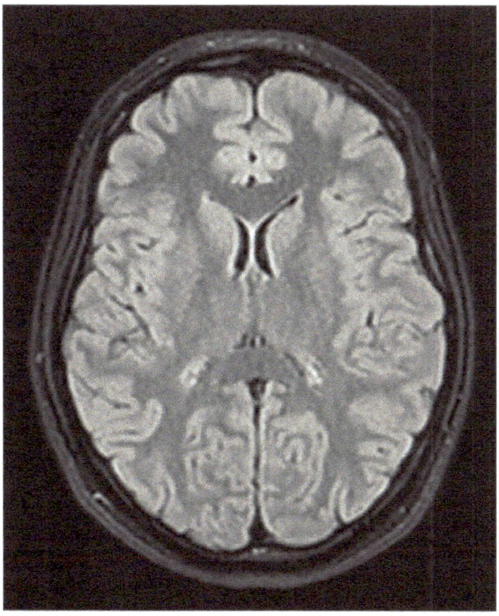

FIG. 9.8 T2 FLAIR brain image. *(MRI scans, courtesy of the Mediskan, Zawiszy Czarnego 7A, 40–872 Katowice.)*

cerebrospinal fluid. It mainly concerns brain imaging. The TI time used in the range from 2000 to 2500ms in combination with a long TR of about 8000ms and a long TE of 140ms gives a T2-weighted signal with the cerebrospinal fluid signal attenuation. The pathological changes, which can be detected in this way, are demyelinating diseases, edema, and ischemic changes [12] (Fig. 9.8).

Gradient Echo Sequence

It is initiated by a single RF pulse and at the same time a gradient is used, directed in the opposite direction, coding the frequency of protons precession. The RF pulse flips the longitudinal magnetization by an angle smaller than 90 degrees; this angle is referred to as Flip angle and lies between 10 and 80 degrees, allowing a faster regrowth of longitudinal magnetization. Depending on the angle measurements, the images may be T1weighted at a larger angle and T2 weighted at a smaller angle. It is a method that requires shorter TR and TE times [2, 7].

The Contrast Agents Used in MRI

As contrast agents in MRI, substances classified into two categories are used:

- paramagnets, which include gadolinium and iron. Their task is to amplify the signal due to decreasing the longitudinal relaxation T1;
- superparamagnetic nanoparticles, which include iron oxides, affecting the transverse relaxation T2. Theoretically, they are less harmful compared to the contrasts used in CT [13].

3D RECONSTRUCTIONS IN CT AND MR EXAMINATIONS APPLIED IN MODELING FOR 3D PRINTING

3D printing is currently used in the diagnostics and surgical procedures planning, primarily in cancer, but also in diseases and injuries of the osteoarticular system. In a large number of cases, 3D printing is used in the reconstruction of bone components, primarily in dentistry, orthopedics, and traumatology. The use of images from multidetector tomography provides volume data allowing for obtaining a virtual 3D model. Bone structures are characterized by a good contrasting, which allows the use of an easy image segmentation process using simple automatic algorithms. Generating a 3D model is crucial in preparing a 3D printout. Parenchymal organs are usually less contrasting and cannot be separated so easily from neighboring organs; hence, automatic segmentation algorithms are not always effective. Therefore, the applied image segmentation procedures can be:

- fully automatic, using an algorithm that automatically divides the image into areas with similar characteristics, for example, the degree of X-ray absorption, and those with characteristics different from the one set;
- semiautomatic algorithm, using a segmentation algorithm but requiring manual change in the points where low contrasted structures are in contact;
- manual segmentation with the outline of structures being marked by the observer;

After extracting the structure and generating the model, it can be changed according to the needs of the planned surgery. 3D printouts are used primarily in planning surgical procedures of the bone and joint system as well as the development of access routes to the operated structures and testing of possible stabilization methods to be used during the procedure. They are also used to create individually matched implants or orthoses. In skull surgeries, 3D printouts are used primarily in planning reconstructive procedures after extensive injuries and cancerous changes [14].

3D printing has been used in the planning of facial reconstruction surgery to treat an orbital blowout fracture. Based on CT studies using MIMICS software, a virtual 3D model of bone part requiring transplantation was prepared. The model was converted to the STL (Standard Triangulation Language) format and a real model of ABS filaments was created on a 3D printer. Thanks to the obtained 3D model, a bone template was created, which was to be implanted in a patient, this time made of aluminum. During the procedure, the surgeon could exactly match the initial template to the place where it was to be implanted. Preparation of an individual model of the planned bone implant allowed for reduction of the final time of surgery. In addition, a more accurate reconstruction of bone structures allows for restoration of the anatomical position of the orbital soft tissues, preventing possible cases of ischemia due to the pressure by dislocated bone fragments and reducing the complications associated with it [15].

Similar application 3D printing is applied in the reconstruction of the sacral bone after its hemisacrectomy during osteosarcoma surgery. To design the bone implant, it is necessary to use the MR examination, performed before the planned operation and the MIMICS software. The rendering of the sacrum is carried out using automatic segmentation. 3D model achieved was fully individualized and adapted to neighboring constructions, and moreover with space for fastening elements. After designing the 3D implant, it is printed on a 3D printer and placed on the site of the removed fragment of the sacrum with a tumor. It is possible to eliminate bone defects resulting from inaccurate implant fitting and to avoid consequences such as pain, experienced by the patient, resulting from limited stabilization of the sacral bone. In the course of a conventional operation, further corrective surgeries are necessary in order to maximally match the implant to the surrounding structures. Thanks to the applied method, the pain complaints resulting from the limited stability of the pelvic area could be reduced and the number of corrective procedures is limited. Stabilization of the sacroiliac joint area allowed for faster rehabilitation and walk, with the use of ankle-stabilizing orthosis, despite nerve damage on one side [16].

3D printing was proposed as a modern method facilitating the diagnosis of pancreatic cancer. Based on multidetector computed tomography, the mass of the tumor can be accurately extracted to plan the future resection procedure. The disadvantage of this imaging method is the appropriate visualization of only those areas of the gland that have good vascularity. To differentiate between cancerous and healthy tissue and to plan the scope of the procedure, a specific algorithm is used to allow for semiautomatic segmentation of pancreatic tissue based on the hypodense nature of its signal and then corrected manually by a radiology specialist. Obtained 3D virtual model includes both the organ itself and vessels adjacent to the pancreas and allows for 3D printing. The printed model allows for precise location of the tumor and vessels with which it is adjacent and other abdominal organs. Although not reproducing the tissue structure, it does not provide the possibility of developing a full course of surgery but is definitely helpful in imaging the precise anatomy of the tumor [16].

REFERENCES

[1] Lusik H, Grinstaff MW. X-ray computed tomography contrast agents. Chem Rev 2013;113(3):1641–66.

[2] Siczek M. Tomografia Komputerowa i Rezonans Magnetyczny dla studentów kierunku Informatyka. Lublin: Instytut Informatyki UMCS; 2011. Esus Agencja Reklamowo–Wydawnicza Tomasz Przybylak.

[3] Prokop M, Galanski M. In: Gołębiowski M, Bekiesińska-Figatowska M, Bestry I, Pasowicz M, Tarasów E, editors. Spiralna i wielorzędowa tomografia komputerowa człowieka. Warszawa: Medipage; 2015. p. 2–35. red. wyd. pol..

[4] Budzik G, Dziubek T, Turek P. Podstawowe czynniki wpływające na jakość obrazów tomograficznych. Basic factors affects on a quality of tomographic images. Problemy Nauk Stosowanych 2015;3:77–84.

[5] Budzik G, Turek P. Proces rekonstrukcji obrazów tomograficznych. Reconstruction process of tomography images. Problemy Nauk Stosowanych 2016;4:57–64.
[6] Maher MM, Kalra MK, Sahani DV, Perumpillichira JJ, Rizzo S, Saini S, Mueller PR. Techniques, clinical applications and limitations of 3D reconstruction in CT of the abdomen. Korean J Radiol 2004;5(1):55–67.
[7] Pruszyński B. Diagnostyka obrazowa – Podstawy teoretyczne i metodyka badań. Warszawa: PZWL; 2014.
[8] Schild HH. MRI made easy. Berlin: Schering AG; 1990.
[9] Walker MT, Partovi S, Karis JP, Fram EK. Fast, versatile, and cost-effective FSE MR imaging: technical considerations and clinical applications. Barrow Quarterly 2000;16:1–5.
[10] Sugimoto H, Sakai O, Shinozaki T, Ohsawa T, Kimura T. Effect of water fraction in selection of optimal TI value for STIR sequences. J Comput Assist Tomogr 1994;18(1):119–25.
[11] Hajnal JV, Bryant DJ, Kasuboski L, Pattany PM, De Coene B, Lewis PD, Pennock JM, Oatridge A, Young IR, Bydder GM. Use of fluid attenuated inversion recovery (FLAIR) pulse sequences in MRI of the brain. J Comput Assist Tomogr 1992;16(6):841–4.
[12] Shokrollahi H. Contrast agents for MRI. Korean J Couns Psychother 2013;33(8):4485–97.
[13] Auricchio F, Marconi S. 3D printing: clinical applications in orthopaedics and traumatology. EFORT Open Rev 2016;1:121–7.
[14] Fan B, Han C, Ying-Jian S, Bei-Fen W, Lin C, Shu-an L, Guang-Yu L. Clinical effects of 3D printing-assisted personalized reconstructive surgery for blowout orbital fractures. Graefes Arch Clin Exp Ophthalmol 2017;255:20151–2057.
[15] Kim D, Lim J, Shim K, Han J, Yi S, Yoon D, Kim KN, Ha Y, Ji GY, Shin DA. Sacral reconstruction with a 3D-printed implant after Hemisacrectomy in patient with sacral osteosarcoma: 1-year follow-up result. Yonsei Med J 2017;58(2):453–7.
[16] Marconi S, Pugliese L, Del Chiaro M, Pozzi Mucelli R, Auricchio F, Pietrabissa A. An innovative strategy for the identification and 3D reconstruction of pancreatic cancer from CT images. Updates Surg 2016;68:273–8.

FURTHER READING

Jones J, Bashir U. K space, https://radiopaedia.org/articles/k-space.

Chapter 10

Quality Control of Biomaterials—Overview of the Relevant Technologies

Joanna Gola
Department of Molecular Biology, School of Pharmacy with the Division of Laboratory Medicine in Sosnowiec, Medical University of Silesia, Katowice, Poland

INTRODUCTION

Any biomaterial produced or modified under laboratory conditions can potentially cause a number of side effects. The occurrence of adverse effects depends not only on the type of biomaterial but also on the method of processing, modification, manufacturing process, sterilization, and the place of implantation [1]. Therefore, it is necessary to carry out the biocompatibility assessment known as "the ability of a material to perform with an appropriate host response in a specific application" [2]. It is recommended to test not only the sample of the material but also the medical device, since the residuals that are not present in the biomaterials produced in the laboratory may be present in the final product. The biocompatibility assessment includes in vitro and in vivo tests and is regulated by International Standard ISO-10993 [3].

Another issue is the determination of the biological activity of scaffolds, which are deliberately enriched by certain factors (e.g., antibiotics, anticancer agents, anti-inflammatory drugs, or growth factors) necessary for cell colonization (e.g., fibroblasts) or homing/differentiation of stem cells. Such modified biomaterials should be assessed not only for biocompatibility but also for bioactivity, both in terms of bioavailability of the biologically active agent and the fulfillment of the planned function, as well as the potential for adverse effects [1]. In the case of biomaterials derived from natural plant and animal tissues, assessment of the acellularization process' efficiency is extremely important, because an immune reaction and the transmission of pathogens may occur [4]. Biomaterial can also be enriched with gene vectors (gene-activated matrix) to provide stable transfection of cells seeded on its surface [5]. Such modifications can cause severe side effects, including immune response and danger of

Stem Cells and Biomaterials for Regenerative Medicine. https://doi.org/10.1016/B978-0-12-812258-7.00010-1

mutagenesis and carcinogenesis. Therefore, the biocompatibility assessment should include not only standard ISO tests but also a number of other studies aimed at identifying the safety of modern tissue regeneration methods, especially in the development of new bioactive materials.

BIOCOMPATIBILITY EVALUATION OF MEDICAL DEVICES

Preliminary Remarks

Biological assessment is important in identifying the potential risks arising from not only the properties of the biomaterial but also from the entire manufacturing process of the medical device by which the final product may acquire new features compared to the starting material [3]. Any modification in the manufacturing process or change of biomaterial supplier may also cause an incompatibility. Such situations, for example, may be the result of incomplete rinsing of toxic solvents. Prior to biological evaluation, potential risks should be estimated based on an exhaustive review of the publicly available information (incl. Information about devices previously reviewed by Food and Drug Administration and The European Commission), clinical and animal study experience, and medical device standards. Risk assessment is necessary to avoid mistakes and knowledge gaps in planning workflows in biocompatibility tests, including physicochemical properties and preclinical biological evaluation.

Extraction techniques are very helpful in the identification of toxic compounds of a medical device prior to biological tests. They are also helpful in the identification of intermediate and final degradation or breakdown products (in the assessment of biodegradable or in situ polymerizing materials). In turn, exhaustive extraction is helpful in a long-term toxicity assessment, especially in the identification of carcinogens. Such an extraction is carried out until dilution of the extractable is 10% of its initial concentration (assessed by gravimetric analysis) [3]. However, submicron or nanotechnology components may require special conditions for extraction, as well as additional evaluation (e.g., absorption, distribution in the organism, accumulation into particular organs, metabolism, and elimination/detoxification). Chemical toxicity is not the only important issue: surface properties, geometry, presence of particulates, and other physical properties (e.g., mechanical forces) can cause incompatibility of medical devices. In addition, mechanical failure of the medical device can alter the biological response because of changes of the physicochemical parameters. Therefore, the biomaterial must be fully characterized in terms of chemical and physical properties. To support a biocompatibility assessment, preliminary information for each material regarding the following should be included: name and formulation (incl. Chemical abstract services (CAS) number, function, structure, and weight percent of each component), intended use, processing methods (incl. Information about additives and impurities), method and conditions of sterilization process, methods used for chemical characterization (due to ISO 10993-18), methods used for characterization of physicochemical, morphological, and topographical properties (ISO 10993-19), and results of chemical and physical properties assessment essential for biocompatibility screening studies.

Planning Biocompatibility Assessment

Even meticulous physicochemical analysis is not sufficient to assess biocompatibility due to the many interactions that can occur in living cells and organisms. Therefore, the next step is to perform an in vitro and in vivo evaluation. Depending on the type of the biomaterial and its destination (location) and time of exposure after implantation, as well as the biodegradation time (if any), specific tests are required. First, in vitro assessment is performed, which limits the number of tests on animals (in vivo). These biomaterials, which in the in vitro tests did not show toxicity, are then assigned to further biocompatibility evaluation steps.

Due to the nature of contact with the body, medical devices are categorized as the following:

1. Surface device—intact skin; mucosal membrane; breached or compromised surface.
2. External communicating device—blood path, indirect contact; tissue (include tissue fluids and subcutaneous spaces), bone, dentin; circulating blood.
3. Implant device—tissue (include tissue fluids and subcutaneous spaces), bone; blood [3].

Further division takes into account contact duration of medical devices: limited—up to 24h; prolonged—from 24h to 30days; and permanent—over 30days [3]. All tests should be performed according to good laboratory practice.

Basic tests valid for all medical devices, regardless of the nature of body contact and its duration, include assessment of cytotoxicity, sensitization, and irritation or intracutaneous reactivity [3]. Depending on medical device category, additional tests may be necessary to assess the following: acute systemic toxicity, material-mediated pyrogenicity, subacute/subchronic toxicity, genotoxicity, implantation, hemocompatibility, carcinogenicity, reproductive/developmental toxicity, and biodegradation of biomaterial (Table 10.1). For many specialized materials, these tests may be insufficient, and in such cases, it is necessary to extend these mentioned tests with neurotoxicity, immunotoxicity, or other specific target organ/tissue toxicity evaluations.

In this part, only tests that are obligatory for all medical devices will be discussed.

Cytotoxicity Tests

Cytotoxicity assays are necessary for the assessment and characterization of potentially toxic and harmful effects of a biomaterial's compounds [4]. The following are cell-culture-based tests recommended by ISO 10993-5: direct contact test, indirect contact test, and extract test (Fig. 10.1) [6]. In all tests, the incubation time should not be shorter than 24h.

TABLE 10.1 Biological Effect Assessment Depending on Medical Device Category, Recommended by ISO and/or FDA

Biological Effect	ISO 10993 Part	Category of Medical Device
Acute systemic toxicity	11	All devices, except contact with skin (regardless of contact duration) and mucosal membrane (limited duration time)
Material-mediated pyrogenicity	11	All devices, except contact with skin (regardless of contact duration) and mucosal membrane (limited duration time)
Subacute/ subchronic toxicity	11	All devices, except contact with skin (regardless of contact duration) and all devices of limited contact duration
Genotoxicity	3	• Permanent duration contact with: mucosal membrane, breached or compromised surface, blood path • Prolonged and permanent contact with tissue, bone, dentin • All devices, which contact with blood
Implantation	6	• Prolonged and permanent contact with mucosal membrane, breached or compromised surface, tissue, bone, dentin, circulating blood • Blood path with permanent indirect contact • All implants contacting with blood
Hemocompatibility	4	External communicating devices (except tissue, bone, dentin); implant devices which contact with blood
Chronic toxicity	11	All devices with permanent contact, except intact skin contact
Carcinogenicity	3	All devices with permanent contact, except intact skin and mucosal membrane contact
Reproductive/ developmental toxicity	3	Novel materials, materials with a known reproductive or developmental toxicity, devices with relevant target populations (e.g., pregnant women), and/or devices where there is the probability for local presence of device materials in the reproductive organs
Degradation	9, 13, 14, 15, 16	Any devices, device components, or materials remaining in contact with tissue that is intended to degrade

Detailed information can be found in the particular parts of ISO 10993.
Based on Use of International Standard ISO 10993-1. Biological evaluation of medical devices—Part 1: Evaluation and testing within a risk management process. Guidance for industry and Food and Drug Administration staff. U.S. Department of Health and Human Services Food and Drug Administration Center for Devices and Radiological Health; 2016.

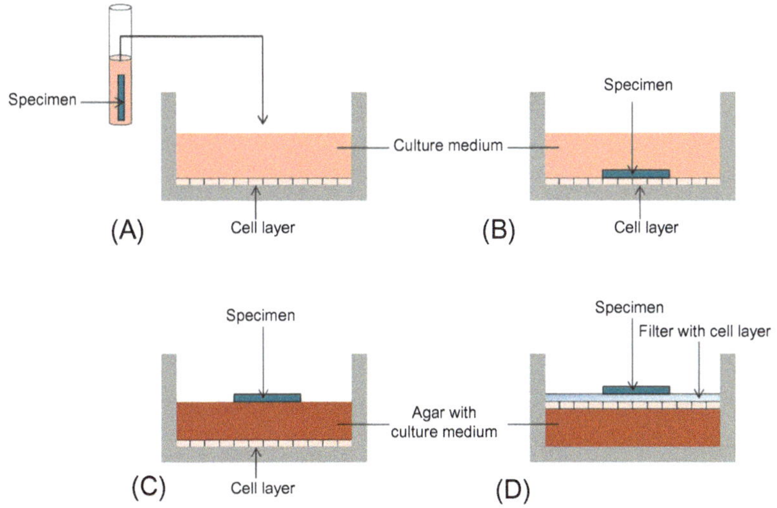

FIG. 10.1 Cell-culture based tests for cytotoxicity assessment of biomaterial specimen. (A) Extract test; (B) direct contact test; (C) indirect contact test: agar diffusion assay; (D) indirect contact test: filter diffusion assay.

In the extract test, cytotoxicity is assessed qualitatively (microscopic observation of cell morphology and reactivity zone) and quantitatively (measurement of cell viability). Liquid extracts of material should be prepared according to ISO 10993-12 [6]. For in vitro tests, extracts may be prepared in culture medium with serum (preferred), physiological saline solution, purified water, and dimethyl sulfoxide (DMSO) (Fig. 10.1A). It should be noted that due to cytotoxic properties, the volume fraction of extract prepared in DMSO should not be greater than 0.5% of culture medium during the cytotoxicity test. The measurement of cell viability could be performed according to internationally validated protocols: neutral red uptake (NRU) cytotoxicity test and colony formation cytotoxicity test or MTT (3-(4,5-dimethylthiazol-2-yl)-2,5-diphenyl tetrazolium bromide) and XTT (sodium 3′-(1-(phenylamino)-carbony)-3,4-tetrazolium)-bis(4-methoxy-6-nitro)benzene-sulfonic acid hydrate) cytotoxicity tests (also recommended by ISO 10993-5). With respect to the principles of good laboratory practice and intra- and interlaboratory precision, researchers may also apply other tests to measure cell death, inhibition of cell growth and proliferation, and cellular metabolism (e.g., release of enzymes, vital dye, and reduction of vital dye) [6].

In contact tests, a qualitative assessment of cytotoxicity is made by microscope observation of changes in cell morphology and reactivity zones. The direct contact test involves placing the specimen of the biomaterial on the cell layer in the center of the vessel (Fig. 10.1B). Agar diffusion or filter diffusion assays are indirect contact tests. For the agar diffusion test, cells are seeded in the vessels, and after growing to subconfluency, they are covered with the mixture of culture medium and melted agar (Fig. 10.1C). After solidification of the agar layer, specimens of the sample material are placed on its surface. The filter diffusion test involves seeding of cells on the filter surface (Fig. 10.1D). After growing to subconfluency, the filter covered with cells is placed on the solidified agar layer (containing culture medium) cell side down. Then, specimens of the biomaterial are placed on the top side of the filter.

Sensitization Evaluation

Biomaterial leachables and extractables can cause skin sensitization, or skin-, eye-, and mucosal irritation [3]. Internationally validated and accepted sensitization assessment involves in vivo animal tests [3]. Detailed recommendations for the implementation of these tests are available in OECD Guidelines for Testing of Chemicals: guinea pig maximization test (GPMT) and closed-patch test (Buehler test) [7], the local lymph node assay (LLNA) [8]. In guinea pig tests, skin reaction after exposure to chemical by epidermal application and or by intradermal injection is observed. The LLNA test is performed on mice. Sensitizers induce lymphocyte proliferation in the local lymph node, which is evaluated by incorporated radioactivity measurements after termination of the animals. Modifications to the LLNA test have been developed to avoid the need for radioactive labeling in the measurement of lymphocyte proliferation [8, 9].

To reduce the number of animal tests, in vitro and in chemico pretests can be performed. So far these methods have not been approved and accepted as standard procedures, but progress has been made in the development of such assays. These tests are based on scientific knowledge concerning chemical interactions at the molecular level resulting in biological responses responsible for adverse effects. The sequence of events creating the adverse outcome pathway (AOP) for skin sensitization is as follows:

1. Covalent binding of chemicals (or their metabolites/abiotic transformation products) to cysteine and/or lysine residues of skin proteins;
2. Inflammatory response in keratinocyte (changes in the expression of genes related to particular pathways);
3. Activation of dendritic cells with the expression of specific proteins;
4. T-cell proliferation in the local lymph node (Fig. 10.2) [10].

The first stage of the process is bioavailability of the chemical that is defined as the ability to penetrate the *stratum corneum*. Available methods for the assessment of percutaneous absorption are in vivo and in vitro tests [10, 11]. The skin of guinea pig, pig, and monkey has the same permeability as human skin, whereas the skin of rats and rabbits is more permeable. For the in vivo method, the most commonly used species is rat, while in vitro models use excised skin from mammals, including humans. In both methods, it is recommended that the tested chemical is radiolabeled. Apart from the marking problem, these tests do not always reflect the real conditions [10].

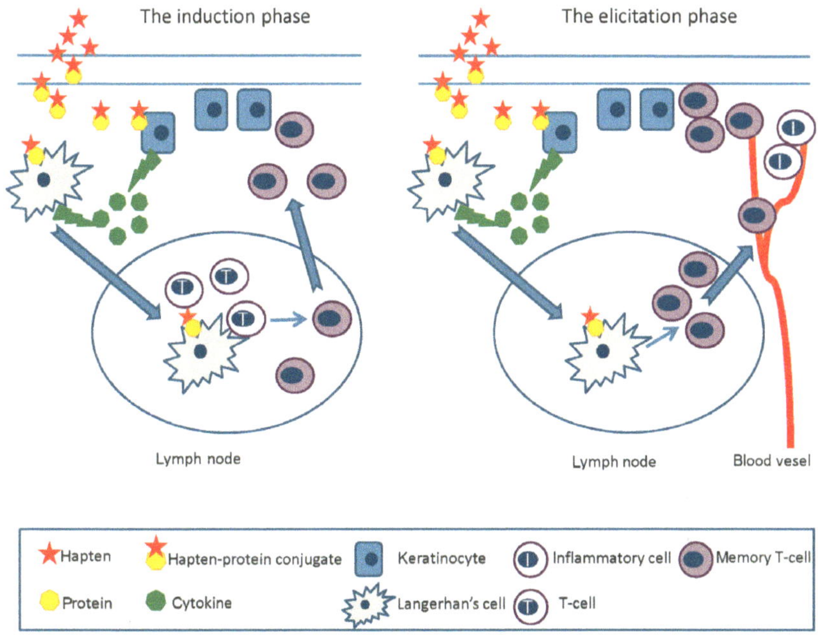

FIG. 10.2 Adverse outcome pathway (AOP) for skin sensitization. The induction phase events are penetration of chemical (hapten) of the outer epidermis, formation of hapten-protein conjugates, secretion of cytokines, interaction with Langerhan's cells and presentation of the antigen to T-cell, differentiation and proliferation of memory t-cells. The elicitation phase occurs after another contact with the same allergen and leads to secretion of cytokines and recruitment of inflammatory cells from circulation. *(Based on OECD. The adverse outcome pathway for skin sensitisation initiated by covalent binding to proteins. OECD Publishing; 2014.)*

Covalent binding to proteins, the first event of AOP, can be assessed by the in chemico method—direct peptide reactivity assay (DPRA) [12]. In this test, synthetic peptides containing cysteine or lysine are incubated with the chemical for 24 h. For detection purposes, they additionally contain phenylalanine. The remaining concentration of peptides is measured using high-performance liquid chromatography (HPLC) with gradient elution and a UV detector with a 220-nm signal. More structurally complex chemicals can be unpredictable in terms of the potential for multiple reactions, despite in silico models [10]. Thus, the suitability of such tests in evaluating the biocompatibility of biomaterials (especially biodegradable ones), in which the extracts are usually a mixture of compounds and their metabolites, seems to be limited.

In recent years, in vitro sensitization tests have been developed. OECD recommends the ARE-Nrf2 luciferase method (antioxidant response element—nuclear factor erythroid 2 [NFE2]-related factor 2) and human cell line activation test (h-CLAT) [12, 13]. These assays address the second and third key events of the adverse outcome pathway in skin sensitization. In the ARE-Nrf2 luciferase method, a transgenic immortalized cell line of human keratinocites (HaCaT) is used. The constitutive promoter of the luciferase gene is fused with the ARE element of the Nrf2-dependent inducible gene (e.g., *AKR1C2* [aldo-keto reductase family 1 member C2] in KeratinoSens assay). After a reaction of Keap1 (Kelch-Like ECH Associated Protein 1) with the sensitizer, the transcriptional factor Nrf2 translocates to the nucleus and binds to the ARE in the promoter of the Nrf2-dependent gene [14]. In the ARE-Nrf2 luciferase test, transcriptional activation of the Nrf2-dependent gene is measured by luminescence detection [13]. In the h-CLAT method, changes in the expression of specific cell surface markers on a human monocytic leukemia cell line (THP-1) are measured by flow cytometry [15]. These markers (e.g., CD86 or CD54) are associated with the activation of monocytes and dendritic cells after exposure to sensitizing chemicals.

In vitro and in chemico tests, along with in silico methods, can be helpful in the sensitization assessment of biomaterials, since positive results of such strategies may abrogate further tests on animals. However, it must be taken into account that negative results do not fully exclude sensitization properties of tested materials. In such a situation, animal tests should be performed.

Irritation/Intracutaneous Reactivity Evaluation

For the assessment of skin, eye, and mucosal irritation, internationally validated and accepted tests include both in vitro and in vivo methods. In vivo assessment involves the acute dermal irritation/corrosion [16] and acute eye irritation/corrosion [17] tests on albino rabbits. There are also numerous in vitro tests for the identification of skin or ocular irritants and corrosives that use cell, tissue, or synthetic models (Table 10.2).

BIOAVAILABILITY AND BIOACTIVITY ASSESSMENT

Preliminary Remarks

The role of biomaterial is related not only to the supplementation of tissue defects but also in the regeneration such tissue. Therefore, the developed biomaterials are enriched by a number of bioactive factors that support this process by providing structural and environmental conditions for the residing cells [28]. One of the most common modifications of biomaterials is their enrichment with growth factors [29]. Biomaterials can also be enriched with other factors, such as anti-inflammatory, antimicrobial, or anticancer drugs. The presence of biologically active substances in the biomaterials dictates not only biocompatibility evaluation but also bioavailability, bioactivity, and the risk of additional side effects assessment, especially in terms of biodegradable material. Biologically active factors can potentially react with the decomposition products of biodegradable biomaterial, and the products of this interaction can cause a number of side effects. Therefore, it is necessary to rigorously define the biological activity and toxicity of not only the starting material but also decay products in terms of

TABLE 10.2 In Vitro Tests for the Assessment of Skin or Eye Corrosion or Irritation

Name of the Test	Model	Main Principle	Refs.
In vitro skin corrosion: transcutaneous electrical resistance test method (TER)	Skin disc taken from killed rats	Loss of normal *stratum corneum* integrity and barrier function of skin discs, measured by a reduction in the transcutaneous electrical resistance	[18]
In vitro skin corrosion: reconstructed human *epidermis* (RHE) test method	Multilayered model of the human epidermis derived from human epidermal keratinocytes	The penetration of corrosive chemicals through the stratum corneum measured by cell viability of underlying layers (based on ability to enzymatic conversion of the MTT vital dye)	[19]
In vitro skin irritation: reconstructed human *epidermis* test method		The penetration of chemicals through the multilayered model measured by the ability of living cells to enzymatic conversion of the MTT dye	[20]
In vitro 3T3 NRU phototoxicity test	Balb/c 3T3 cells monolayer	Loss of cell viability of mouse fibroblasts exposed to chemical in the presence of nontoxic irradiation, measured by the uptake and binding of Neutral Red vital dye	[21]
In vitro membrane barrier test method for skin corrosion	Synthetic macromolecular biobarrier	Damage of membrane barrier by corrosive chemicals, measured by the color change of pH indicator or combination of dyes in the presence of chemical	[22]
Fluorescein leakage test method	Cell monolayer that models in vivo corneal epithelium	Loss of tight junctions and desmosomal junctions of tubular epithelial cells (MDCK CB997) monolayer grown on semipermeable inserts, measured by the passage of sodium-fluorescein dye through the monolayer and insert membrane.	[23]

(Continued)

TABLE 10.2 In Vitro Tests for the Assessment of Skin or Eye Corrosion or Irritation—cont'd

Name of the Test	Model	Main Principle	Refs.
Reconstructed human cornea-like epithelium (RhCE) test method	Three-dimensional RhCE tissue constructs derived from primary human cells that models in vivo corneal epithelium	The penetration of chemicals through three-dimensional RhCE tissue construct measured by the ability of living cells to enzymatic conversion of the MTT dye	[24]
Short time exposure in vitro test method	Statens seruminstitut rabbit cornea (SIRC) cells monolayer	Loss of cell viability of SIRC cells measured by the ability of living cells to enzymatic conversion of the MTT dye	[25]
Bovine corneal opacity and permeability test method	Corneas isolated from the eyes of freshly slaughtered cattle	Changes in corneal opacity and permeability measured with the use of an opacitometer and a visible light spectrophotometer	[26]
Isolated chicken eye test method	Eye collected from freshly slaughtered chicken	Damage of the eye evaluated based on corneal swelling, opacity, and fluorescein retention	[27]

their possible interactions with other components. Predicting and modeling interactions of degradation products with biologically active compounds and their metabolites is very important. In such cases, in vitro tests are not sufficient to investigate all possible interactions. In vivo studies can demonstrate that products of these interactions do not show a direct threat (by toxicity), but this risk may arise from the generation of a microenvironment favorable for the development of inflammation or cancer.

Bioavailability

Bioavailability/release of biologically active agents depends not only on the material itself and the method of binding/closing agents in the material but also upon the environmental conditions to which the factor is released [29]. Some strategies use environmental proteins, such as metalloproteinases, to release agents from the scaffold. Depending on the purpose of use, biologically active agents can be applied in gradient concentration, in different spatial systems, and at different densities to control processes such as adhesion, migration, or stem cell differentiation. The closing/loading rate of the biologically active agent in the material can be assessed using, for example, the dynamic light scattering (DLS) technique, scanning electron microscopy (SEM), energy-dispersive X-ray spectroscopy (EDX), Fourier transform infrared spectroscopy (FTIR), or by using an appropriate method to evaluate the concentration of the agent in the extract [30]. Bioavailability assessment involves determining the rate of release and the concentration of the released factor. For this purpose, a series of extracts (at different incubation times, even several dozen days) are performed in which the concentration of the biologically active agent is determined [30]. Depending on the type of biologically active agent, various methods can be used to determine its concentration in the extract. For example, for the assessment of growth factors, commonly used methods are enzyme-linked immunosorbent assay (ELISA), UV–visible spectroscopy, or reverse phase HPLC (RP-HPLC) [31]. In the case of drugs, many methods are available, depending on their chemical structure, that not only assess the concentration of the drug and its metabolites but also detect any impurities and products of interaction with other compounds (e.g., biomaterial degradation products) [32]. Some such analytical techniques are titrimetric and chromatographic techniques, spectroscopic techniques, electrochemical and electrophoretic methods, and kinetic methods [32]. The release of ions of silica, calcium, and anions of oxidated potassium may be evaluated by inductive coupled plasma atomic emission spectrometry (ICP-AES) [33].

The bioavailability assessment may also include the determination of a factor in the tissue surrounding the site of scaffold implantation. In the case of a biologically active agent, it should be taken into account that its concentration may be overestimated due to its natural occurrence in the tissue to which it is released [1, 28]. It was shown that biomaterials are not completely inert and can stimulate the surrounding tissues to secrete such factors [1, 28]. For example, crosslinked chitosan may stimulate the surrounding cells to secrete cytokines such as the tumor necrosis factor (TNF), while crosslinked hyaluronic acid gels stimulate secretion of tissue plasminogen activator (tPA) [1]. In tissue regeneration, the role of TNF is primarily related to its proinflammatory activity, which may lead to the development of chronic inflammation [1]. On the other hand, induction of inflammation is one of the elements of tissue regeneration [34]. Another problem may be the limitation of the bioavailability of the released growth factor due to the masking of the domains responsible for interactions with receptor proteins through which the agent manifests its activity [29]. Thus, in the next stage, the bioactivity of the factor used should be assessed.

Bioactivity Assessment

The bioactivity evaluation can be carried out by means of in vitro tests (at the cellular or subcellular-molecular level) or by tests in animals (in vivo). In vitro assays at cell level include the assessment of changes in morphological features and cell activity (the ability to proliferate, migrate, adhere, or differentiate).

At the subcellular level, changes in intracellular signaling, activation of transcription factors, changes in gene expression (mRNA and protein levels), interactions, and metabolic pathways are evaluated.

Biological assays can be performed using extracts of the material (Fig. 10.3). The extract is added to the cell culture, and, after a certain period of time, the expected biological effect is evaluated. Tests are selected according to the expected biological activity of the factor. For example, bioactivity of Nell-1 (Nel-like molecule-1) promoting osteochondrocyte function is assessed based on the activity of alkaline phosphatase (ALP) in the osteosarcoma cell line [30]. ALP is also used as an indicator of early interim osteoblast differentiation [35]. The immunomodulatory effect of biomaterials with anticancer vaccines can be assessed based on their potential to promote infiltration and trafficking of immune cells [36]. Antitumor effects of biomaterials are usually evaluated based on cancer cells' viability and proliferation with the use of MTT or cell counting kit-8 (CCK-8), for example [37–39]. Recently, 3D in vitro tumor models have been developed to provide more information on microenvironment and tumor heterogeneity influence on anticancer drug activity [39–41]. Efficacy of immobilized antimicrobial agents can be evaluated by numerous microbiological methods (e.g., zone inhibition, inoculation [immersive, direct and aerosol], and luminescence methods [microscopy, flow cytometry]) [42].

Another strategy is to evaluate the biological effect on cells homing the scaffold (Fig. 10.3). The assessment of the bioactivity of scaffolds seeded with cells

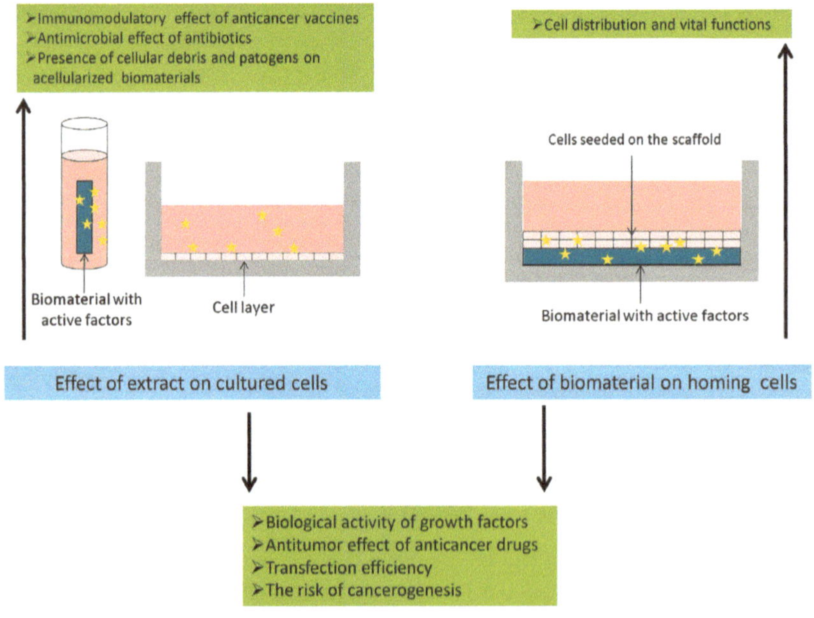

FIG. 10.3 The main aspects of bioactivity evaluation of biomaterials enriched with specific bioactive compounds.

includes, first of all, an assessment of the process of cell implantation and the maintenance of their vital functions, followed by an assessment of the biological activity of the scaffold. Cell distribution and morphology may be evaluated based on the observation of the cytoskeleton and/or nucleus staining via fluorescence, scanning, or electron or confocal laser light microscope [29, 30]. Cell viability and proliferation may be measured with the use of a CCK-8 kit [30] or MTT assay [35]. Cell adhesion could be evaluated using SEM [30]. Regarding the assessment of the influence of growth factors on cells seeded on the scaffold, the evaluation strategy depends on the intended use of tissue construct. Most of tests are used to assess the potential of mesenchymal stem cells (MSC) to differentiate into individual cell types. The ability of Nell-1 to induce chondrogenic differentiation of human bone mesenchymal stem cells (BMSC) may be assessed based on transcriptional activity of *COLII* (*collagen II*), *Aggrecan*, and *SOX9* (*SRY-box 9*) genes, the presence of glycosaminoglycan (GAG) content, histological staining, and collagen I (COLI) and COLII immunohistochemical staining [30]. In turn, BMP-2 (bone morphogenetic protein 2) induced osteoblastic cell line differentiation may be evaluated based on the expression of specific genes (*OPN—osteopontin*, *Runx2—runt related transcription factor 2*) and calcium mineralization (based on alizarin red staining) [35]. Differentiation of BMSC into osteoblasts could be assessed based on gene expression (*BMP-2*, *TGF-β—transforming growth factor-beta*, *OPN*, *ALP*, *COLI*), osteopontin immunohistochemical staining, and alizarin red staining [43].

Performing in vitro pretests allows the rejection of these scaffolds, which causes severe side effects before they are directed to in vivo testing [44]. However, due to the complexity of the healing and regeneration process, in vitro tests do not allow accurate evaluation of tissue regeneration [44]. The great advantage of in vivo testing is the ability to observe the regeneration process with all the side effects that may occur after implantation. On the other hand, these tests require a lot of time, labor, and cost, and despite humanitarian attitudes to animals, there is a lot of ethical controversy. One of the in vivo models used as a pretest to evaluate tissue-engineered grafts is the chick embryo chorioallantoic membrane (CAM) assay [44]. This model is mainly used to study biological processes and in recent years to evaluate biomaterials. One of the advantages of CAM is the lack of immunocompetence, which allows the study of tissue-engineered constructs with human cells. In spite of the numerous references regarding the use of CAM in the assessment of biocompatibility and bioactivity of tissue-engineered constructs, biological differences between mammals and avians should be taken into account.

Acellularized Scaffolds

Another aspect of bioactivity concerns the concept of the implantation of acellularized biomaterials, which are typically cell-free protein matrix constructs (ECM—extracellular matrix) [4]. The acellularization process requires a number of methods, mainly chemical ones, and as a result, the biomaterial may contain potentially toxic molecules. Such materials are designed for host cell infiltration or colonization with stem/progenitor cells [45]. Therefore, any residue of the chemicals may prevent the recellularization process and cause severe side effects [4]. Moreover, almost every agent used for decellularization (physical, biological, or chemical) may damage ECM ultrastructure and affect mechanical properties of the scaffold. In addition, the assessment of the efficiency of the decellularization process is extremely important because of possible immune reaction and the risk of transmission of pathogens [4]. The sources of immunogenicity are DNA remnants and alpha-Gal epitope (α-Gal) expressed in most mammalian tissues excluding humans and primates. It is believed that the amount of DNA remnants should be less than 50 ng dsDNA per mg of dry ECM. This could be evaluated with the use of common methods, like spectrophotometric quantification or UV fluorescence labeling (e.g., PicoGreen, DAPI—4′,6-diamidino-2-phenylindole). Gel electrophoresis is also usable because DNA fragments shorter than 200 bp in length do not cause inflammatory reactions in vivo [4]. The presence of α-Gal can be confirmed by immunohistochemistry and ELISA methods. To exclude the inflammatory response, in vitro and in vivo tests can be performed. In vitro tests usually determine several cell functions and involve macrophage-, monocyte-, and lymphocyte-based assays [4]. To assess the possible transmission of pathogens, several methods can be used, including real-time qPCR (quantitative polymerase chain reaction). This technique was performed to detect complete degradation of porcine endogenous retrovirus (PERV) DNA in porcine-derived heart valves [46]. The research showed that depending on the method and time of the decellularization process, PERV subtypes may be still present in the biomaterial. Moreover, in some cases, porcine DNA of the *GAPDH* (glyceraldehyde-3-phosphate dehydrogenase) gene was also detected, indicating the presence of debris of porcine cells. Other methods used for pathogen detection involve hybridization-based assays and several biosensors, including a recently developed liquid-crystal-based DNA biosensor [47].

Gene Activated Matrix

In addition to the MSC differentiation strategies under the influence of growth factors, methods for the transfection of cells seeded on scaffolds have been developed [5]. If this process is performed at in vitro conditions, both viral and nonviral vectors can be used. However, the usefulness of viral vectors for in vivo transfection is under question because of possible severe side effects that involve danger of mutagenesis and immune response. Thus, the most preferable strategy for in vivo gene delivery is nonviral vectors such as the following: lipidic carriers, polymer base systems, cell-penetrating peptides, and inorganic materials (e.g., hydroxyapatite). Constant release of nucleic acids from a biomaterial enriched with gene vectors—a gene activated matrix—provides efficient and sustained transfection of cells seeded on its surface [5]. One of the most interesting solutions for efficient nucleic acid delivery is magnetic transfection complexes consisting of paramagnetic iron oxide nanoparticles coated with streptavidin interacting with biotin labeled microRNA and polyethylenimine complexes [48].

Gene activated matrix bioactivity should be evaluated in terms of cytotoxicity (which may be quite high for lipid-based vectors) and transfection efficiency [5]. Cytotoxicity can be evaluated using the aforementioned methods, such as cytotoxicity assays utilizing tetrazolium salt (MTT, XTT). Transfection efficiency can be assessed by the expression of the therapeutic and/or reporter gene. The ability of cells to capture a transgene and its expression can be evaluated at the RNA (using real-time RT-qPCR—reverse transcriptase qPCR) and protein level (by ELISA assay, immunocytochemistry, Western blot). Use of reporter gene (e.g., coding green fluorescent protein, GFP) allows direct observation of transfected cells by means of fluorescence microscopy [49]. Use of such a marker for the cells seeded on the scaffold greatly facilitates the evaluation of transfection efficiency.

The Risk of Cancerogenesis

Factors and signaling pathways involved in tissue regeneration are also related to the development of a cancer, which can be seen as a consequence of impaired regeneration [34]. Many growth factors used for the development of bioactive scaffolds for tissue regeneration [29] have been reported as molecules that promote cancer growth; for instance, VEGF (vascular endothelial growth factor) and bFGF (basic fibroblast growth factor) are essential for cancer angiogenesis [50], and BMP-2 is involved in the differentiation of cancer stem cells [51]. Biomaterials may also change cellular metabolism; for example, acrylate monomers deplete glutathione, which is one of the major antioxidants [52]. This situation leads to an increase of reactive oxygen species (ROS) and an increase in genotoxic effects. Moreover, several biomaterials may affect biological and molecular processes, including intracellular signaling, cell cycle regulation, and cell differentiation [53].

The risk of cancerogenesis also arises from the fact that tumor stem cells may be present in every tissue and under favorable conditions will initiate tumor growth [54]. Cancerogenesis depends not only on the molecular changes of cancer cells but also on the tumor microenvironment. Intercellular interactions and paracrine signals lead to changes in normal stromal cells (fibroblasts, immune and endothelial cells) present in the surrounding of the tumor. Molecules secreted by normal cells (chemokines and growth factors), together with ECM, create a specific microenvironment promoting tumor growth and progression [54]. The presence of scaffold-derived growth factors could facilitate environmental changes leading to cancerogenesis.

To better understand stages of cancer progression, in vitro 3D models have been developed [54]. At the same time, these studies show that the structure of the scaffold also affects the promotion of cancer. If a tumor is initiated during tissue regeneration, the scaffold itself can further support this process. It was reported that large collagen fibers in collagen gels promote invasiveness of breast cancer cells, while small collagen fibers hamper their migration [55]. Angiogenesis was promoted by 3D hyaluronan hydrogels, which enhanced the expression of $VEGF_{165}$ and interleukin-8 [56]. Together, with deliberately introduced growth factors, biomaterials may sustain or promote cancerogenesis at the site of implantation. Some of these factors may act as chemoattractants for circulating cancer cells. Moreover, 3D models showed that some solid biomaterials (e.g., bone models) enhance cancer metastasis [54]. On the other hand, biomaterials may be used to design modern therapy in which the tumor microenvironment is modified [54].

Scaffold colonization with differentiated cells or stem cells also generates the risk of tumor development. Carbon nanotubes and metal-based nanoparticles have been reported to increase ROS production, alter DNA integrity, and influence cell life functions [57]. This can consequently lead to the development of the cancer process. Stem cells possess the ability of unlimited proliferation, and, together with mutations and the presence of growth factors, they may transform into cancer stem cells [58]. The strategy of cell reprogramming also increases the risk of cancerogenesis due to genomic instability and the mutation rate [53]. Another problem may arise from incomplete differentiation of cells, transgenic reactivation, and epigenetic modifications [59].

In the assessment of the risk of cancer development, genetic toxicity testing according to ISO 10993-3 is necessary [60]. Recommended tests are the bacterial reverse mutation test [61] and mammalian cell genotoxicity tests [62–65]. The use of tumor cell cultures in bioactivity studies further increases the probability of detecting undesirable effects of biomaterial. Additionally, 3D models may be very helpful in the detection of biomaterial's potential to promote the development of cancer and its metastasis [54]. To overcome the problem of cell transformation into cancer cells, undifferentiated cells can be eliminated by sorting (magnetic- or fluorescence-activated) or by cytotoxic antibodies, prior to scaffold seeding [59]. There are also several methods available for the detection of cancer cells and cancer stem cells [66].

CONCLUDING REMARKS

Biomaterials and tissue-engineered constructs bring biotechnologic options to regenerative medicine. However, several aspects should be considered before launching such solutions. Meticulous planning of biocompatibility and bioactivity assessments, together with conscientious research and critical interpretation of the results, allows for the selection of optimal solutions that will not pose a risk to the patient.

REFERENCES

[1] Kohane DS, Langer R. Biocompatibility and drug delivery systems. Chem Sci 2010;1(4):441–6.
[2] Williams DF. On the mechanisms of biocompatibility. Biomaterials 2008;29(20):2941–53.
[3] Use of International Standard ISO 10993-1. Biological evaluation of medical devices—Part 1: Evaluation and testing within a risk management process. In: Guidance for industry and Food and Drug Administration staff. U.S. Department of Health and Human Services Food and Drug Administration Center for Devices and Radiological Health; 2016.
[4] Hussein KH, Park K-M, Kang K-S, Woo H-M. Biocompatibility evaluation of tissue-engineered decellularized scaffolds for biomedical application. Mater Sci Eng C 2016;67:766–78.
[5] Raisin S, Belamie E, Morille M. Non-viral gene activated matrices for mesenchymal stem cells based tissue engineering of bone and cartilage. Biomaterials 2016;104:223–37.
[6] Biological evaluation of medical devices. In: Part 5: Tests for in vitro cytotoxicity. ISO; 2009.
[7] OECD. Test No. 406: skin sensitisation. OECD Publishing; 1992.
[8] OECD. Test No. 429: skin sensitisation. OECD Publishing; 2010.
[9] OECD. Test No. 442A: skin sensitization. OECD Publishing; 2010.
[10] OECD. The adverse outcome pathway for skin sensitisation initiated by covalent binding to proteins. OECD Publishing; 2014.
[11] OECD. Test No. 427: skin absorption: in vivo method. OECD Publishing; 2004.
[12] OECD. Test No. 442C: in chemico skin sensitisation. OECD Publishing; 2015.
[13] OECD. Test No. 442D: in vitro skin sensitisation. OECD Publishing; 2015.
[14] Jaramillo MC, Zhang DD. The emerging role of the Nrf2-Keap1 signaling pathway in cancer. Genes Dev 2013;27(20):2179–91.
[15] OECD. Test No. 442E: in vitro skin sensitisation. OECD Publishing; 2016.
[16] OECD. Test No. 404: acute dermal irritation/corrosion. OECD Publishing; 2015.
[17] OECD. Test No. 405: acute eye irritation/corrosion. OECD Publishing; 2012.
[18] OECD. Test No. 430: in vitro skin corrosion: transcutaneous electrical resistance test method (TER). OECD Publishing; 2015.
[19] OECD. Test No. 431: in vitro skin corrosion: reconstructed human epidermis (RHE) test method. OECD Publishing; 2016.
[20] OECD. Test no. 439: in vitro skin irritation: reconstructed human epidermis test method. OECD Publishing; 2015.
[21] OECD. Test No. 432: in vitro 3T3 NRU phototoxicity test. OECD Publishing; 2004.
[22] OECD. Test No. 435: in vitro membrane barrier test method for skin corrosion. OECD Publishing; 2015.
[23] OECD. Test No. 460: fluorescein leakage test method for identifying ocular corrosives and severe irritants. OECD Publishing; 2012.
[24] OECD. Test No. 492: reconstructed human cornea-like epithelium (RhCE) test method for identifying chemicals not requiring classification and labelling for eye irritation or serious eye damage. OECD Publishing; 2015.
[25] OECD. Test No. 491: short time exposure in vitro test method for identifying (i) chemicals inducing serious eye damage and (ii) chemicals not requiring classification for eye irritation or serious eye damage. OECD Publishing; 2015.
[26] OECD. Test No. 437: bovine corneal opacity and permeability test method for identifying (i) chemicals inducing serious eye damage and (ii) chemicals not requiring classification for eye irritation or serious eye damage. OECD Publishing; 2013.
[27] OECD. Test No. 438: isolated chicken eye test method for identifying ocular corrosives and severe irritants. OECD Publishing; 2009.

[28] Stratton S, Shelke NB, Hoshino K, Rudraiah S, Kumbar SG. Bioactive polymeric scaffolds for tissue engineering. Bioact Mater 2016;1(2):93–108.
[29] Reed S, Wu B. Sustained growth factor delivery in tissue engineering applications. Ann Biomed Eng 2014;42(7):1528–36.
[30] Wang C, et al. Two-phase electrospinning to incorporate growth factors loaded chitosan nanoparticles into electrospun fibrous scaffolds for bioactivity retention and cartilage regeneration. Mater Sci Eng C 2017;79:507–15.
[31] Quarmby V, Quan C, Ling V, Compton P, Canova-Davis E. How much insulin-like growth factor I (IGF-I) circulates? Impact of standardization on IGF-I assay accuracy. J Clin Endocrinol Metab 1998;83(4):1211–6.
[32] Siddiqui MR, AlOthman ZA, Rahman N. Analytical techniques in pharmaceutical analysis: a review. Arab J Chem 2017;10:S1409–21.
[33] Lv F, et al. A conducive bioceramic/polymer composite biomaterial for diabetic wound healing. Acta Biomater 2017;60:128–143.
[34] Brockmann L, Giannou A, Gagliani N, Huber S. Regulation of TH17 cells and associated cytokines in wound healing, tissue regeneration, and carcinogenesis. Int J Mol Sci 2017;18(5):1033.
[35] Zhao X, et al. BMP-2 immobilized PLGA/hydroxyapatite fibrous scaffold via polydopamine stimulates osteoblast growth. Mater Sci Eng C 2017;78:658–66.
[36] Bencherif SA, et al. Injectable cryogel-based whole-cell cancer vaccines. Nat Commun 2015;6:7556.
[37] Kandaswamy N, Raveendiran N. Synthesis, characterization, and in vitro antimicrobial and anticancer evaluation of copolyester bearing 4-arylidene curcumin in the main chain. Int Scholar Res Notices 2014;2014:1–6.
[38] Gakhar G, et al. Anti-tumor effect of novel cationic biomaterials in prostate cancer. Anticancer Res 2014;34(8):3981–9.
[39] Liang J, Sun XS, Yang Z, Cao S. Anticancer drug camptothecin test in 3D hydrogel networks with HeLa cells. Sci Rep 2017;7:37626.
[40] Gu L, Mooney DJ. Biomaterials and emerging anticancer therapeutics: engineering the microenvironment. Nat Rev Cancer 2015;16(1):56–66.
[41] Girard YK, et al. A 3D fibrous scaffold inducing tumoroids: a platform for anticancer drug development. PLoS ONE 2013;8(10):e75345.
[42] Green J-BD, Fulghum T, Nordhaus MA. A review of immobilized antimicrobial agents and methods for testing. Biointerphases 2011;6(4):MR13–28.
[43] Tian Y, et al. Osteoblast-oriented differentiation of BMSCs by co-culturing with composite scaffolds constructed using silicon-substituted calcium phosphate, autogenous fine particulate bone powder and alginate *in vitro*. Oncotarget 2017;8(51):88308–88319.
[44] Baiguera S, Macchiarini P, Ribatti D. Chorioallantoic membrane for in vivo investigation of tissue-engineered construct biocompatibility. J Biomed Mater Res B Appl Biomater 2012;100B(5):1425–34.
[45] Aibibu D, Hild M, Wöltje M, Cherif C. Textile cell-free scaffolds for in situ tissue engineering applications. J Mater Sci Mater Med 2016;27(3):63.
[46] Cyganek-Niemiec A, et al. Degradation effect of diepoxide fixation on porcine endogenous retrovirus DNA in heart valves: molecular aspects. Int J Artif Organs 2012;35(1):25–33.
[47] Khan M, Khan AR, Shin J-H, Park S-Y. A liquid-crystal-based DNA biosensor for pathogen detection. Sci Rep 2016;6(1).
[48] Schade A, et al. Innovative strategy for microRNA delivery in human mesenchymal stem cells via magnetic nanoparticles. Int J Mol Sci 2013;14(6):10710–26.

[49] Atikah AN, et al. Identification of transfection efficiency using qualitative and quantitative analyses of green fluorescent protein in CHO cells. J Appl Sci 2014;14(5):489–92.
[50] Rajabi M, Mousa S. The role of angiogenesis in cancer treatment. Biomedicine 2017;5(2):34.
[51] Carreira ACO, Zambuzzi WF, Rossi MC, Filho RA, Sogayar MC, Granjeiro JM. Bone morphogenetic proteins. In: Vitamins & hormones. vol. 99. Elsevier; 2015. p. 293–322.
[52] Schmalz G, Galler KM. Biocompatibility of biomaterials—lessons learned and considerations for the design of novel materials. Dent Mater 2017;33(4):382–93.
[53] Wasik AM, et al. Reprogramming and carcinogenesis—parallels and distinctions. In: International review of cell and molecular biology. vol. 308. Elsevier; 2014. p. 167–203.
[54] Adjei I, Blanka S. Modulation of the tumor microenvironment for cancer treatment: a biomaterials approach. J Funct Biomater 2015;6(1):81–103.
[55] Carey SP, Kraning-Rush CM, Williams RM, Reinhart-King CA. Biophysical control of invasive tumor cell behavior by extracellular matrix microarchitecture. Biomaterials 2012;33(16):4157–65.
[56] Fischbach C, Kong HJ, Hsiong SX, Evangelista MB, Yuen W, Mooney DJ. Cancer cell angiogenic capability is regulated by 3D culture and integrin engagement. Proc Natl Acad Sci 2009;106(2):399–404.
[57] Hofmann M-C. Stem cells and nanomaterials. In: Capco DG, Chen Y, editors. Nanomaterial, Vol. 811. Dordrecht: Springer Netherlands; 2014. p. 255–75.
[58] Hombach-Klonisch S, et al. Cancer stem cells as targets for cancer therapy: selected cancers as examples. Arch Immunol Ther Exp (Warsz) 2008;56(3):165–80.
[59] Cieślar-Pobuda A, et al. Transdifferentiation and reprogramming: overview of the processes, their similarities and differences. Biochim Biophys Acta Mol Cell Res 2017;1864(7):1359–69.
[60] Przygoda RT. Safety assessment and global regulatory requirements for genetic toxicity evaluations of medical devices: genetic toxicity evaluations of medical devices. Environ Mol Mutagen 2017;58(5):375–9.
[61] OECD. Test No. 471: bacterial reverse mutation test. OECD Publishing; 1997.
[62] OECD. Test No. 473: in vitro mammalian chromosomal aberration test. OECD Publishing; 2016.
[63] OECD. Test no. 490: in vitro mammalian cell gene mutation tests using the thymidine kinase gene. OECD Publishing; 2016.
[64] OECD. Test no. 476: in vitro mammalian cell gene mutation tests using the Hprt and xprt genes. OECD Publishing; 2016.
[65] OECD. Test No. 487: in vitro mammalian cell micronucleus test. OECD Publishing; 2016.
[66] Akbari-Birgani S, et al. Cancer stem cells, cancer-initiating cells and methods for their detection. Drug Discov Today 2016;21(5):836–42.

FURTHER READING

OECD. Test No. 442B: skin sensitization. OECD Publishing; 2010.
OECD. Test No. 428: skin absorption: in vitro method. OECD Publishing; 2004.

Chapter 11

Methods of Biomaterial-Aided Cell or Drug Delivery: Extracellular Matrix Proteins as Biomaterials

Batoryna Olgierd, Aleksandra Sklarek, Paulina Siwek, Ewa Waluga
Department of Community Pharmacy, Faculty of Pharmacy with the Division of Laboratory Medicine in Sosnowiec, Medical University of Silesia, Katowice, Poland

INTRODUCTION

Currently, there can be observed an increasing interest in using natural materials both in regenerative medicine and drug delivery systems. A broad scope of applications in this aspect is mainly offered by a family of collagen protein and glycosaminoglycans (GAGs) [1, 2]. A great interest in raw materials is linked to the need of obtaining a safe product biocompatible with body tissues [3]. Despite numerous scientific studies, conducted on collagen proteins and GAGs, there is still a constant search for new solutions, which would enable the development of new products with desired physiochemical and pharmacological properties.

COLLAGEN PROTEINS—CHARACTERISTICS

Collagen proteins belong to the most widespread proteins in the organism of mammals and additionally constitute 1/3 of all proteins occurring in vertebrates [4, 5]. They are the main component of the extracellular matrix (ECM) of connective tissue forming many organs. The tissue is a building block of skin, ligaments, bones, tendons, and cornea.

The majority of the collagen particle occurs in the form of an α-superhelix built of three mutually twisted polypeptide chains. The structure is the result of the amino acid composition of the collagen particle. In terms of the structure, the feature distinguishing the collagen protein molecules from the other proteins of the organism is the presence of the repeated sequence -Gly-X-Y- in which the role of glycine residues is to maintain tightly packed collagen chains,

Stem Cells and Biomaterials for Regenerative Medicine. https://doi.org/10.1016/B978-0-12-812258-7.00011-3

while X and Y are the residues of proline and hydroxyproline, respectively [6]. Proline and hydroxyproline, possessing a heterocyclic structure, contain a nitrogen atom in the ring, which influences the tightening of the polypeptide bond. The spiral-shaped collagen chain results in this case from the amino acid residues being repeated in the triplet sequence of the polypeptide chain [6].

According to the available data, currently there are 29 genetically separate types of collagen proteins. These differences arise from a various length of the protein molecule's helical part. Moreover, gaps in the triple-helix structure, size and shapes of globular domains as well as changes in posttranslational modifications additionally affect this variety [7]. Collagen types: I, II, III, V, and XI are specific fibrous proteins, collagens types IX and XII belong to the proteins bound to fibrils, while collagens type IV and VII create a network structure [8].

The nomenclature of collagen proteins is not fully systematized and is most commonly used for numbering particular types of collagens with Roman figures, while α chains being their parts are marked with Arabic numerals, e.g., α_1(I).

Collagen proteins play different functions in the organism. They are mainly responsible for biochemical properties of tissues and organs constituting their stability and strength [8]. Additionally, they support cells' growth, influence their migration, morphology, and adhesion as well as differentiation. Collagen types I, II, III perform a role in tissue reconstruction by revealing a chemotactic action [9].

COLLAGEN PROTEINS—SYNTHESIS AND ASSEMBLY

Biosynthesis of collagen protein covers the common processes for the biosynthesis of all proteins as well as specific for collagen synthesis. The process of collagen biosynthesis consists of transcription and translation of polypeptide chains of preprocollagen, which are the products of single genes and are synthesized in whole. The next stage is the signal peptide cleavage. Then, the proline residues hydroxylation occurs, specific only for collagen proteins, in the process of which some proline residues are hydroxylased in the 3 or 4 position. This reaction is catalyzed by proline-4-hydroxylase and proline-3-hydroxylase and requires cofactors in the form of ascorbate and ferrous ions. Moreover, it is indispensable for creating a helical structure [10].

Subsequently, the specific stage for collagen proteins biosynthesis is hydroxylation of lysine residues, catalyzed by lysine hydroxylation. The transition facilitates formation of crosslinks and embedding carbohydrates into the synthesized protein molecule [11]. Next, the previous synthesized hydroxyproline and hydroxylysine residues are exposed to hydroxylysine galactosylotransferase as well as to galactosylo-hydroxylysin glucosyltransferase. This transition decreases the collagen protein susceptibility to collagenase [10].

In nonhelical parts of the procollagen particle, there occurs a glycosylation of asparagine residues. Then, three procollagen chains conjoin in the shape of a three-wire rope, stabilized by formation of disulfide bonds, which is catalyzed by protein disulfide-isomerase. The procollagen molecules are transformed into the collagen molecules after an enzymatic cleavage of the amino- and carboxyl-terminal part as a result of the reaction catalyzed by specific for the collagen type and the terminal part of peptidase [2].

Next, in the process of collagen biosynthesis, there occurs formation of various supramolecular structures depending on the protein type. The necessary stage for obtaining appropriate stability of collagen fibrils is the development of crosslinks in joining the polypeptide chains within the same as well as between the adjacent molecules. This process mainly involves lysine oxidase dependent on cupric ions [2, 12].

Finally, the created collagen structure interacts with proper elements of the connective tissue, such as GAGs, proteoglycans (PGs), and other structural glycoproteins in order to provide optimal biomechanics properties of tissues and organs [12].

COLLAGEN PROTEIN DEGRADATION

Collagen is degraded extracellularly and intracellularly. Extracellular degradation of collagen is based on depolymerization of supramolecular structures, action of tissue collagenase, denaturation of collagen fragments at body temperature, and further decomposition of collagen by nonspecific proteases. Intracellular degradation however comprises the action of collagenolytic cathepsins active in an acidic environment [13].

Collagen fibrils organized in a triple-helical structure are resistant to proteolytic enyzmes' action. However, single-stranded regions of the collagen protein molecule are susceptible to matrix metalloproteinases (MMPs). Particularly, types I, II, III of collagen are hydrolyzed by such collagenases as MMP-1, MMP-2, MMP-8, MMP-13, and MMP-14 [14]. The created fragments of collagen fibrils are further decomposed by gelatinases and nonspecific proteases. Other MMPs, such as MMP-3 and MMP-9, possess the ability to bind with collagen type I but do not participate in its degradation [14]. The collagenolytic activity of the mentioned MMPs is dependent on the ability to bind collagen protein molecule, and unravel α-helix chain of collagen as well as the ability to degrade single polypeptide chains [13, 14].

THE METHOD OF OBTAINING COLLAGEN PROTEINS AS BIOMATERIALS

Due to a widespread distribution of collagen proteins in mammals' organisms, e.g., in skin or tendons, the organs are particularly often used as a primary source of collagen for medical purposes (as implants, dressings, drug delivery systems). Additional group of raw materials to be obtained from collagen proteins covers the connective tissue of human placenta as well as human collagen obtained from genetic recombination in transgenic organisms. The properties of obtained proteins are influenced not only by insolubility in organic solvents and water but also by the animal's age and the type of the obtained tissue [8].

Basically, four types of collagen proteins can be isolated from tissues and purified and therefore used in the pharmaceutical industry as drug delivery systems. These types include natural collagen soluble in salts, collagen treated enzymatically and alkaline, collagen soluble in acids, and also insoluble collagen [15]. A great strength of collagen proteins and their resistance to proteolytic enzymes' action results mainly from their occurrence in the covalent crosslinks structure.

COLLAGEN PROTEINS AS BIOMATERIALS—ADVANTAGES

In comparison with other proteins, collagens belong to the group of proteins with weaker immunogenic properties. The main antigenic determinants are localized in telepeptide regions of the collagen molecule and can be eliminated by pepsin. The next two types of determinants consist of a triple-helix and amino acid sequence occurring in the alpha conformation, while the last one occurs only for denatured proteins [16]. Availability and a variety of collagen sources make the protein particularly suitable for the usage as a biomaterial.

COLLAGEN PROTEINS AS BIOMATERIALS—DISADVANTAGES

The fundamental disadvantage of collagen biomaterials is their gradual calcification occurring in body tissues [17]. The calcification of the collagen molecule is dependent on two mechanisms. The first of them causes formation and deposition of calcium crystals, phosphate ions between the protein fibrils. The second mechanism is about the formation of vesicles (by budding of the cell membrane), in which calcium and phosphate ions collect. The level of calcification of the collagen fibrils must be strictly control in the biomaterials applied in soft tissue (tendons, ligaments, skin, heart valves, pericardium, etc.). The collagen biomaterials, which are slightly calcified, undergo a fast decomposition after implantation. Collagen intermolecular crosslinking formation with glutaraldehyde (GTA) favors fast dystrophic calcification of the protein after the implant placement. The process of calcification may be inhibited by application of surfactants (sodium dodecyl sulfate, diphosphonates, dimethylosulfoxide or metal ions, or by such crosslinking agents as acyl azide or polyepoxides [18, 19]). Another disadvantage in the common application of collagen in the pharmaceutical industry is its price [20].

COLLAGEN PROTEINS AS BIOMATERIALS—MODIFICATIONS

The crosslinking of collagen in vitro by physical processing or chemical agents modify its physicochemical properties decreasing solubility, water absorption ability, and influencing biodegradability of the collagen biomaterial. Moreover, it changes its mechanical properties.

The antigenicity of collagen biomaterials may be modified by the crosslinking process, which could be divided into three main types: chemical crosslinking, physical crosslinking, and enzymatic crosslinking [16, 17]. The crosslinking agents used in the chemical crosslinking are among others formaldehyde, carbodiimides, acyl azides, and polyepoxides. Despite the fact that this type of crosslinking leads to a significant increase in the stability of the biomaterial, too high concentration of the crosslinking agents, resulting from the polymer material degradation, may contribute to occurrence of the cytotoxic effects. Another type of a chemical stabilization of collagen is to modify the amino groups of the polypeptide chains on polycation molecules such as chitosan. Due to the application of such a solution, there is a considerable decrease of the cytotoxic potential of the biomaterial [18, 19].

In the process of physical crosslinking UV and gamma radiation, the process of dehydrothermal (DHT) treatment crosslinking could be applied. The exposure to the mentioned agents contributes to an increased thermal and proteolytic resistance of the collagen material. It is assumed that the UV radiation with the wavelength of 254 nm increases the resistance of collagen to proteases' action, while the DHT treatment leads to a decrease in its sensitivity to trypsin and susceptibility to being degraded by pepsin and collagenolytic cathepsin. The application of the gamma radiation as a crosslinking agent requires water due to the fact that in a dry environment the polypeptide chains crack [16, 17]. The application of such agents as the UV or gamma radiation as well as DHT minimizes the risk of cytotoxicity after implanting the material. The UV radiation causes free radicals to be generated, which in turn are concentrated to the greatest extent on the core of the amino acid residues, such as tyrosine and phenylalanine. Together with the radiation energy, the crosslinking level of collagen increases. The gamma radiation causes gaps in the polypeptide chains, while in the presence of water the intermolecular crosslinks are formed. The DHT procedure links to a thermal processing of collagen for several days. It causes an increase in the temperature of protein denaturation by removing water from its molecule, after which the intermolecular crosslinking phase occurs as a result of the condensation reactions such as esterification or formation of an amide binding. This reaction involves the residues of aspartic acid, glutamic acid, serine, threonine, arginine, and lysine [18]. The advantage of the collagen crosslinking by physical processes is the lack of reagents in the polymer matrix, which are the reason for a potential cytotoxicity of the matrix. However, the long-term use of UV and gamma radiation or increased temperature on the collagen proteins is controversial. Additionally, the collagen material modification by physical agents affecting the stabilization or destabilization of the biomaterial structure is of a dynamic balance character and depends on the proper selection of parameters of the conducted process. The physical agents applied to modify the collagen matrix directly influence the loss of valuable properties of the biopolymer material [18].

The crosslinking of polypeptide chains of collagen with enzymes, especially transglutaminases, makes it possible to obtain a biopolymer material, which while degradation does not generate cytotoxic products [19].

The crosslinking of collagen affects its ability to react with drugs. A non-crosslinked collagen binds drugs with higher capacity when compared to a crosslinked collagen. The more crosslinked structures in turn possess wider possibilities to more permanently "trap" substances [21].

The methods used of altering the polypeptide chains of the collagen biomaterials include the modifications of side chains by means of changing the value of the electrostatic charge by succinylation or methylation. Such modifications may alter the collagen solubility depending on the pH of the environment and interactions of the biomaterial with among others blood platelets. Moreover, such procedures may contribute to activation and increase in proliferation of fibroblasts in vitro.

The change in hydration of collagen leads to formation of interrelations between the amine or carboxyl groups of amino acids during condensation and more specifically the reactions of esterification of hydroxyl and carboxyl groups [22].

The surface modifications of the biomaterials play a significant role in the tissue engineering and regenerative medicine by increasing the ability for cell adhesion and improving the biocompatibility.

A simple ways of adding the ECM components to the surface of the polymer carrier is coating, which is a nonspecific physical surface adsorption of the polymer material. Supplying the immobilized bioactive molecules into the tissue is one of the ways of supporting the tissue regeneration process. An example of drug delivery systems is a dextran-based sirolimus hybrid capsule. The cardiovascular stents produced with the coating method have a great biocompatibility and a favorable release profile. The ECM proteins, such as collagen, may be used to affect the cells' behavior. The collagen IV coatings influence the cell's differentiation toward the components of cardiovascular system. Fibronectin and gelatin, in turn, are applied as a coating to improve cells' adhesion and proliferation of bladder smooth muscles. Coating is an attractive and cost-effective method enabling drug delivery into the surrounding tissues. The coating processes, which can be used just before implantation, are more beneficial that the coating used at an early stage of the production process. It allows the application of specific doses for a given patient, decreases the need to store the products under defined rigorous conditions, especially those having a short half-life [1].

Electrospinning is an attractive method of generating fibrous and porous matrices with high surface/volume ratio for regenerative medicine. The changing properties of the matrix such as the size of fibrils and pores enable the control of the cell response including differentiation, migration, and progenitor cells' proliferation. An example of the electrospinning method application is a collagen mesh formed of collagen type I and polyethylene oxide. Generally, electrospinning is a simple and effective method of generating the materials containing 3D fibrils of such substances as collagen, hyaluronate, PGs, and GAGs [1].

Bioprinting is a computer-controlled technique of an additive production of biomaterial layer by layer. The 3D scaffold may be generated with the use of a laser, or via controlled extrusions through nozzles, while the latter printing method is appropriate for printing cells. It has been proven that the ECM proteins such as collagen, fibrin, gelatin are appropriate for generating a 3D scaffold with an imprint. The printed chondrocytes show a good viability due to which the *bioprinting* method is proper for generating cell carriers in cartilage tissue engineering based on ECM. *Bioprinting* enables the generation of complex and well-defined 3D structures or even complete organs.

The printing and coating method of the *Patterning on Topography* (PoT) type enables the layering of such proteins as fibronectin, laminin, collagen type IV, and other ECM proteins on the printed geometrical surfaces from smooth to topographically complex ones [1].

COLLAGEN PROTEINS IN DRUG DELIVERY TECHNIQUES

Due to unique structural and physicochemical properties of collagen, the protein is used as an excipient in topical dosage forms [23, 24]. Medical substances are implemented and located in the collagen matrix by generating drug-protein interactions. Such interactions include ionic bonds, hydrogen bonds, and covalent bonds. The medical substances may also be physically "trapped" in the structure of the collagen protein [21].

Depending on the collagen properties, especially its structure, level of crosslinking, and hydration, it may form different dosage forms—from liquids, thought semisolids, to solid formulations; the noncrosslinked collagen occurs as a solution, while the crosslinked collagen is in a solid form [25]. As mentioned earlier, chemical and physical methods may be used to introduce an additional crosslinking bond into the collagen matrix. Collagen biomaterials applied in drug delivery techniques occur as solutions, gels, inserts, films, coverings, sponges, nanomolecules, nanospheres, macromolecules, and tablets [26–28].

In topical drugs, collagen and the active substances synergistically affect the surrounding tissues. This solution is most frequently applied in stopping bleeding or supporting the wound healing process [1]. Collagen carriers are moreover widespread in ophthalmology, stomatology, surgery, and oncology, mainly in the form of coverings, inserts, sponges, injections, and films [29, 30]. For films formation, collagen-containing solutions are used.

Suspensions and collagen solutions are used to fill and correct soft tissues. Such preparations often include an additional local anesthetic, e.g., lignocaine. Collagen gels are used for stabilization of dispersion systems. Moreover, they can also be a carrier for active substances, for example, a gel containing 5-fluorouracil and epinephrine used for increasing the effectiveness of the anticancer therapy.

In ophthalmology, collagen films are applicable in the form of inserts or protective membranes. The overall hydrolysis of the collagen protein administered as a film occurs after approximately 6h. In this dosage form, collagen is most frequently used to treat inflammations of the cornea [31]. Collagen coverings, which do not contain an active substance, are applied in the production of contact lenses. Such coverings melt slowly and lubricate the corneal surface protecting it against damages. The properties of collagen coverings may be modified with glutaraldehyde, chromium, or UV lighting [26].

Collagen films may also constitute a coating material for implants placed into an organism. It is relatively frequent to find stents coated with a protective layer of a collagen protein film. The coating enables changes of the surface properties of biomaterials. For instance, in practice, stunts are coated with a biomaterial constituting a mixture of collagen proteins and rapamycin [31].

These collagen coverings gradually release active substances. Saturation of the matrix is used water-soluble bioactive substances, whereas water-insoluble drugs are introduced into the polymer matrix with an incorporation method [23].

Collagen coverings are primarily used for treating local infections and injuries, glaucoma and to achieve immunosuppression. The most frequently used active substances in collagen coverings are antibiotics (gentamicin, vancomycin, tobramycin) as well as trimethoprim, ofloxacin, pilocarpine, heparin, and cyclosporin. Such coverings prolong the active substance, cause an increase of the drug bioavailability, and ensure a higher concentration in the area of action [32, 33]. Application of collagen coverings as carriers of active substances has however some limitations. In ophthalmology, they may impair visual acuity, cause discomfort after application, and may mechanically weaken the carrier material [28].

The alternative for collagen coverings is collasomes—pieces or speck of collagen. Their advantage is a possibility to be applied as a drop. The active substance is suspended in a viscous, liquid carrier. It is practiced to apply collagen specks with cyclosporine [33].

A frequently applied dosage form with collagen is a collagen sponge with a porous structure. The biomaterials are obtained mainly by lyophilization, while the porosity of the collagen matrix depends on the freezing time [27].

Collagen sponges are applied to accelerate wound healing and stop bleeding. They quickly and easily absorb tissue secretions and assure wound adhesion maintaining wound moisture and protecting damaged tissues against additional damages. A gradual biodegradation of the sponges enables a proportional tissue rebuilding. The collagen sponges with an incorporated active substance facilitate a synergistic influence of collagen and the active substance on healing wounds. The application of collagen as a carrier of the active substance makes it possible to obtain a higher concentration in the area of action with a simultaneous low concentration in the serum, which significantly lowers the risk of adverse effects [34]. The collagen sponges release active substances in several phases, since their pores are open and partially closed. The active substance is firstly released from open pores, and next from closed pores. However, partially blocked substance is realized by the collagen fibrous structure in the last phase [27].

Among active substances applied in collagen sponges there are gentamycin, cefotaxime, fusidic acid, clindamycin, or vancomycin [24].

THE PERSPECTIVE OF APPLICATION OF COLLAGEN PROTEINS

The properties of collagen can be modified by adding elastin, fibronectin, GAGs, or synthetically degraded polymers such as polylactide, polyglycolide, copolymers of lactide, and glycolide (PLGA) or ε-caprolactone. Due to such a modification, it is possible to obtain a hybrid material with property dependent on the type and amount of the applied components [27].

Collagen is also a material, which in the future may be widely used in the tissue structure or organ restoration enabling the supply of growth factors and genes into the damaged tissue structures. Despite existing disadvantages of collagen biomaterials, it is difficult to replace it with a synthetic polymer used as a carrier of the active substance.

METHODS OF BIOMATERIAL-AIDED CELL OR DRUG DELIVERY—GLYCOSAMINOGLYCANS AS BIOMATERIALS

GAGs are natural polysaccharides, which occur in all animal tissues. They perform a variety biological functions, i.e., they participate in controlling organogenesis and growth, in the processes of proliferation, differentiation, and adhesion of cells, in translation of signals, in homeostasis, in healing processes control, in regulating prooxidative-antioxidative balance, in lipid metabolism, oncogenesis, in the processes of inhibiting and modifying system's immunity as well as in the course of infectious diseases [34–36]. The participation of GAGs in the mentioned processes is largely the result of the interaction of these heteropolysaccharides with numerous proteins including the components of ECM, proteases, proteases inhibitors, cytokines, growth factors, adhesion molecules, and even the surface receptors of viral cells [35, 37, 38].

GAGs have a character of linear macromolecules built of a variable number of repeating disaccharide units each of which consists of one hexosamine residue, *N*-Acetyl-D-Galactosamine (GalNAc) or *N*-Acetyl-D-Glucosamine (GlcNAc) and one hexuronic acid residues, D-glucuronic (GlcA) or L-iduronic (IdoA), or neutral hexose, D-galactose (Gal) [32, 36]. The structural diversity of GAGs, which results from a variable length of the polysaccharide chains, from a few up to 25,000 disaccharides, from a varied degree and place *N*- and *O*-sulfate of disaccharide units, a varied range of their acetylation, a varied degree of epimerization of GLcA to IdoA, a different charge density of glycols and their varied domain order, different places of occurrence, i.e., in the ECM, on the cell's surface, intracellular, as well as diversity caused by multiple interaction with protein cores, determines their biological functions [35, 39]. The presence of acidic sugar residues in the structure of disaccharide and/or sulfate groups, gives GAGs a character of polyanions [40]. In the system, GAGs occur either in the free form or in the form of covalent links to proteins, the so-called PGs. These hybrid macromolecules, composed of a central core protein, are dominated by the attached glycosaminoglycan chains [41].

DIVISION, SIMILARITIES, AND DIFFERENCES

According to the type of the monosaccharide subunits and the kinds of the glycosidic bonds between them, GAGs are divided into four classes [41]:

- hyaluronic acid (also known as hyaluronan) (HA);
- chondroitin sulfates (CS) and dermatan sulfates (DS);
- heparan sulfates (HS) and heparins (Hep);
- keratan sulfates (KS).

Among these, only HA does not form sulfate salts and as the only one does not covalently bind to proteins and therefore is not included in the composition of the proteoglycan's molecule. However, keratan sulfate as the only one does not contain in the disaccharide composition the hexuronic acid residues, but the residues of galactose in this place [42]. In the case of hyaluronan and chondroitin sulfate, the hexuronic acid is glucuronic acid (GlcA), while in dermatan sulfate, heparin sulfate and heparin occur both as glucuronic and induronic acids [36, 41, 43]. GAGs may additionally be divided into two groups of glycans depending on the kind of hexosamine, which is contained in the disaccharide units' structure. There are GAGs, which comprise glucosamine-containing hyaluronan, heparin sulfate, heparin and keratin sulfate, and galactoseminoglycans, which in turn comprise galactosamine-containing chondroitin sulfate and dermatan sulfate [36, 44]. Particular structures of GAGs are shown in Fig. 11.1.

CHONDROITIN SULFATE AND DERMATAN SULFATE

CS, constituting a quantitative fraction of PGs—glycoconjugates, are formed of repeating disaccharide units [→4GlcAβ1→3GalNAcβ1→], which during postsynthetic modifications undergo a sulfation in different monomeric positions of the subunits, i.e., *N*-Acetyl-D-Galactosamine (GalNAc) or GlcA. The varied way of CS and DS sulfating represents their heterogeneity. It comes down to formation either chondroitin sulfate A, in the disaccharide unit, the *N*-Acetylgalactosamine residue of which is sulfated in the 4 carbon of this monosaccharide or to formation chondroitin sulfate, in the disaccharide unit, in which 6 carbon of GalNAc residue is mainly sulfated [36, 41]. The disaccharides of a different number and a varied placement of the sulfate groups, such as double-sulfated, occur in the central fragments of the polysaccharide CS chain [36]. In the case when one of the sulfated group binds in the 2 carbon position of GlcA, while the second in the 6 position of carbon N-Acetylgalactosamine, is called D disaccharide, while in the case when both sulfate groups bind to the GalNAc

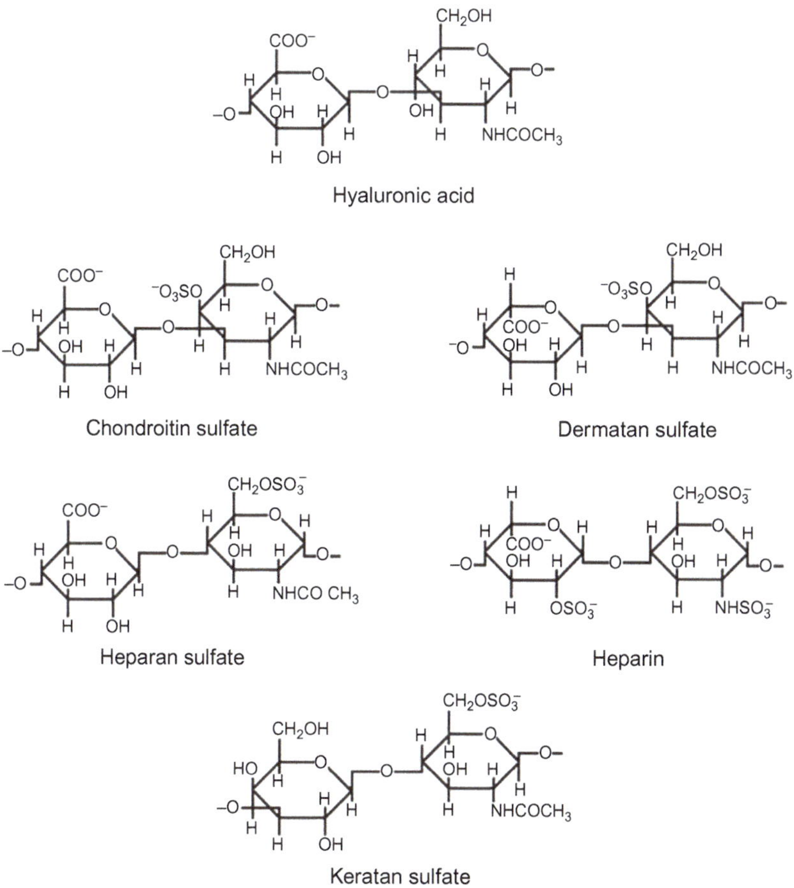

FIG. 11.1 The structure of disaccharide units creating glycosaminoglycans [51].

residue, one in the C-4 position and the second in C-6 position of this monosaccharide subunit, it is called E disaccharide of chondroitin sulfate [36, 41]. The number and position of sulfate groups in the CS chains correlates with the rate of GAGs bioactivity. This is relevant, for instance, in wound healing in the process of binding pathogens or interaction with heparin-binding proteins [45]. The further enzymatic modification of the CS chain includes C-5 epimerization of the D-glucuronic acid into L-iduronic acid, which is often accompanied by sulfating in the C-2 IdoA position, leads to formation of B chondroitin sulfate also known as dermatan sulfate [36, 46]. The polysaccharide DS chain mainly consists of disaccharide units [→4IdoAβ1→3GalNAcβ1→], containing in a smaller proportion double-sulfated units such as those sulfated in the 4 GalNAc carbon and in the 2 IdoA carbon positions. As a result of numerous modification, within the polysaccharide DS chain there occurs highly modified oligosaccharide domains split into regions of a relatively small degree of a structural reconstruction of this GAG [41]. The disaccharide units containing GlcA are mainly sulfated in the position of C-6 *N*-Acetylgalactosamine, while the sulfation in the C-4 position of the mentioned hexosamine is usually associated with the presence of iduronic acid in the structure of this disaccharide [47]. The presence of IdoA in the structure of dermatan sulfate gives this GAG a conformational resilience, seems to be playing a key role in the specificity of the protein-binding site interacting this GAG, and moreover favors the inhibition of fibroblasts' proliferation in a significantly higher degree than the presence in the GlcA glycosaminoglycan chain [38, 41]. The variable length of the polysaccharide DS chain, the variable placement of IdoA in the chain, the variable sulfation of the monosaccharide subunits as well as the diversity of the core proteins, determine the rate of complexity of the DSPG DS-containing PGs [41].

Both dermatan and CS are synthesized as glycosaminoglycan, side chains of PGs are then added. Firstly, the core protein is synthesized, which takes places in the endoplasmic reticulum, and next in the Golgi apparatus monosaccharide residues are attached to the existing proteins [35]. The biosynthesis of the charoitin-dermatan chains is initiated by the activity of xylosyltransferase. This enzyme catalyzes the transfer of the xylose molecule on the serile residue of the core protein, which still takes place in the endoplasmic reticulum. Subsequently, two molecules of galactose are successively attached in the Golgi apparatus to the xylose residue via galactosylotransferase II. The fourth monosaccharide unit attached to the growing sugar fragment is GlcA transferred by glucuronosylotransferase I [46]. These stages of GAGs biosynthesis are common for both chondroitin and DS as well as for heparan sulfates and heparin [46]. The formed tetrasaccharide fragment with the structure [GlcA-Gal-Gal-Xyl], common for CS, DS, and HS/H is called the binding region. The type of glycosaminoglycan, which is to be formed in the process of the further biosynthesis, depends on the enzyme transferring the next sugar molecule [46].

The carbohydrate skeletons of CS/DS and HS/H are only slightly different. Subsequent modifications such as sulfation, epimerization, and deacetylation (occurring only in HS/H) lead to a varied structure of the mentioned GAGs by which they determine their specific biological activity. A great diversity of the chain structure of the compounds in question is the reason that in nature there are no two identical GAGs [46].

The final stage of the CS/DS biosynthesis is IdoA sulfation by 2-*O*-sulfotransferase (CS/DS2ST). The enzyme proliferates the sulfation of the IdoA residues located next to GalNAc residue sulfated in the C-4 position [46].

Numerous modification of CS/DS determines the specific activity of PGs containing in their structure glycan components [39]. A huge number of PGs possess some side CS/DS chains. To the group of the PGs modified by CS chains belong aggrecan, versican, neurocan, brevican, bamacan, or the CD44 isoform [38, 46].

The GAGs of a given type may be covalently attached to different core proteins, which determine their tissue location and biological functions. The expression of GAGs is affected by the cell stimulation with growth factors, which leads to the type change of the GAG binding with the core protein [38]. Moreover, the enzymatic degradation of PGs occurs during a dynamic tissue remodeling, e.g., in the process of wound healing the location and biological activity of these macromolecules is modified. Summing up, the function of PGs in the body is determined by at least three independent systems: the synthesis of the core protein, the synthesis of GAGs, and finally the enzymatic remodeling of the proteoglycan structure [47].

The final stage of the chondroitin-dermatan PGs' modification is their degradation. A partial decomposition of these macromolecules occurs in the extracellular space and takes place with endoglycosidase and proteolytic enzymes. The first ones cover hyaluronidase, which degrades the chondroitin-dermatan glcyans, while the second ones comprise MMPs [48]. The complete degradation of the chondroitin-dermatan PGs occurs only intracellularly, in lysosomes where PGs or their fragments undergo a complete hydrolysis [49].

The biological functions of the chondroitin-dermatan GAGs are determined by their structure modified in the process of the postsynthetic remodeling by the chain size [38, 46]. CSPG and DSPG perform their biological functions by interactions with several ligands [48]. Many interactions with protein ligands occur exclusively with the proteoglycan core protein, in the other only with the glycan component, while in some with both PG components. The interactions reflect one of the most significant functions of PGs, i.e., the input of these compounds in the structural organization of the ECM [50]. Other functions of DS/CSPG are as follows: participation in the intercellular signal transferring, binding and activating or inhibiting the growth factors such as EGF, the family FGF, HGF, SF, TGF-β, interactions with chemokines and cytokines shaping the body immune response, modification of the body immunity to infectious agents, participation in homeostasis, in the body lipid metabolism and in the repair processes of tissue damages and participation of CS/DS in the development of the central nervous system [44, 51].

HEPARAN SULFATES (HS) AND HEPARINS

Heparan sulfates and heparins, the GAGs of the most complex structure of all GAGs, are formed with repeating disaccharide units consisting of the GlcA residues and *N*-acetylgalactosamine with a schematic structure [→4GlcAβ1→4GlcNAcα1→], where the glycoside bond, between the hexuronic acid and *N*-acetylgalactosamin, takes the configuration β1→4 instead of β1→3, while the bond between *N*-acetylgalactosamine and hexuronic acid takes the configuration α1→4 instead of β1→4, as it occurs in the chondroitin-dermatan GAGs [41].

Due to the structure of the same monomeric subunits of heparin and heparan sulfate, both biopolymers are very often taken for analogs, while the range of the postsynthetics modifications, which differs significantly in both cases, considerable distinguishes them [35]. In the case of heparin, the monomeric subunits are sulfated in a significantly bigger extent than the HS subunits [41]. The biggest negative charge density presented by heparin makes it the most acidic macromolecule of the human body [42].

The heparan sulfate molecules are characterized by a greater structural heterogeneity, when compared to the Hep structure. HS contains a bigger amount of the acetylated glucosamine residue compared to *N*-sulfated GlcN, a bigger GlcA contents, while less *O*-binding sulfating groups additionally manifesting a domain structure, covering highly-sulfated, heparin-like sequences, weakly sulfated sequences and totally unmodified regions: [GlcA – GlcNac] [42].

Heparin and heparan sulfate also differ in tissue location, core proteins, to which glycans are bound in the process of biosynthesis as well as the number of glycan chains associated with the protein [41]. Heparin is synthesized in mast cells and basophils in the form of glycosaminoglycan side chains of PGs—serglycin [52]. The synthesis of the heparan sulfate also results from proteoglycan components occurring both on the surface of the cell membrane as well as in the ECM [52].

Initial stages of heparanosulfate PGs' biosynthesis do not differ from the initial stages of CS/DSPG biosynthesis. The tetrasaccharide binding region, associated with the core protein serine residue, initiates the elongation of HS/H chain [52]. In this process, the monosaccharide subunits—*N*-acetylgalactosamine and GlcA bound alternately to nonreducing end of the growing glycan chain, by glucosylotransferase, the so-called exostosin I (EXT1) and exostosin II (EXT2) [46]. During polymerization, the glycan chain undergoes many modifications [52]. They cover *N*-deacetylation and *N*-sulfation of glucosamine, epimerization of GlcA into IdoA, 2-*O*-sulfation of hexuronic acid (usually IdoA), 6-*O*-sulfation and 3-*O*-sulfation glucosamine and start with removing *N*-acetyl groups, and replacing them with sulfation groups catalyzed by an enzymatic complex *N*-deacetylation/*N*-sulfotransferase (NDST). *N*-deacetylation and *N*-sulfation are a precondition for next enzymatic modifications [46]. The subsequent is the transformation of GlcA into iduronic one catalyzed by C-5 epimerase, which is specific for both HS and Hep [52]. Similarly, the activity of 2-*O*-sulfotransferase (HS2ST) presents specificity toward both HS and Hep. The enzyme catalyzes the sulfating reaction of hexuronic acid. The next enzyme in the modification process of the growing HS/Hep chain—6-*O*-sulfotransferase heparansulfate (HS6ST) transfers the 6-*O*-sulfate group on N-sulfated glucosamine, while 3-*O*-sulfotransferase heparansulfate transfers the 3-*O*-sulfate group also on the residue of the mentioned hexaosamine [46].

The modification of the heparansulfate PGs comprise not only the ones associated with the biosynthesis of the glycan chain, but also the modifications concerning the transformation of the core protein of PGs consisting of heparan sulfate proteoglycans (HSPG) [53].

The HSPG cover the family of syndecans, the transmembrane PGs, formed of 4 syndical members 1,-2,-3,-4, sometimes "enriched" with CS chains, the family of glipicans—PGs bound to the cell membrane by glycosylphosphatidylinositol made of 6 members (glipican-1-2-3-4-5-6), betaglycan—PG of the cell surface also known as type III receptor for TGF-β, and moreover—testican (chondroitin-heparansulfate PG of the ECM), or—perlecan and again (PGs of basal membranes) [39].

The manifestation of another postsynthetic modification of HSPG is the release of GAGs chains from their proteoglycan links (the so-called ectodomain shedding), which leads to modification of the nonsoluble protein-associated glycans into soluble, free forms of these compounds. The soluble GAGs may undergo further enzymatic modifications associated with the change of the glycan chain length or with "revealing" the specific, so far "masked" domains enabling a total transformation of the primary function of these biopolymers [35]. A particular role in the process of the enzymatic modifications is played by heparanase (endoglucuronidase)—the HS chain-degrading enzyme [46]. The degradation of heparansulfate PGs initially occurs in the extracellular space, next in lysosomes, according to an analogous scheme for CS/DSPG [46].

The biological functions of heparansulfate GAGs have been well defined. Heparan sulfates play a significant role in the regulation of interactions between cells and between cells and the ECM. HS stimulate the cell adhesion to ECM binding with the matrix macromolecules such as fibronectin and laminin [52]. HS and Hep perform a function of endogenic receptors for numerous extracellular ligands, growth factors, and chemokines regulating the processes of cell proliferation, differentiation and angiogenesis, leucocytal migration and degranulation, but also in cancer progression [54]. The role of HSPG in the metastatic process depends on the tissue type, tumor pathophysiological condition, and tissue location of HSPG. Generally hence not always, the HSPG of the cell surface inhibit metastasis, while the core proteins themselves and HS chains modulate the mentioned process [55]. Heparin synthesized in basophils and mast cells of lungs, intestines, and liver play a role in the body defense mechanisms [42]. It manifests antioxidative, antiinflammatory, and vasodilatory properties [42]. Together with heparan sulfates it shows anticoagulation activity; however, the in vivo function of Hep in the regulation of the coagulation seems to be unlikely. The HSPG of the cell surface regulate the metabolism of lipoproteins, neurogenesis, and tissue damage repair processes [47, 48].

KERATAN SULFATE

Keratan sulfate GAGs occur in animal organisms as proteoglycans (KSPG). They occur in the ECM and on the cell membrane surface [56]. Keratan sulfate GAGs are built of repeating disaccharide units formed of the residues of galactose and *N*-acetyloglucosamine with a schematic structure →3Galβ1→4GlcNAcβ1→] [41]. Postsynthetic modifications of the glycan chain comprise sulfation always in the C-6 position of one or both monomeric subunits, which leads to a formation of mono- or disulfated regions in the glycan chain [57]. The residues of GlcNAc6S within monosulfated KS regions may be subject to fucosylation. Moreover, KS may contain sialic acid residues binding with Gal and Gal6S residues located on a nonreducing chain end of the GAG in question [57].

Keratan sulfate is divided into three types: KS I (corneal), KS II (skeletal), and KS III (cerebral). The basis of the KS division, not the mentioned three types, is the structure of the region binding KS with the core protein [56]. Within KS II—skeletal, there are two subtypes, KS IIA—articular and KS IIB—nonarticular, based on the presence in the first one of α(1–3)fucose residues and the α(2–6)-*N*-acetylneuraminic acid [41].

The biosynthesis of keratin sulfate occurs in two stages. First, the region binding the core protein with GAG is created, while then the elongation of the chain and its modification take place [56]. The elongation of KS chains of all types occurs by alternate attaching of Gal and GlcNAc residues, catalyzed by the activity of β-1,4-galactotransferase and β-1,3-*N*-acetyloglucosaminotransferase, respectively [58]. The modification of KS chains is based on sulfation at C-6 *N*-acetyloglucosamine residues and galactose [41]. Sulfation covers mainly the GlcNAc residues, while to a lesser extent the galactose residues [57]. The sulfation of the hexosamine residues is catalyzed by *N*-acetyloglucosamine-6-*O*-sulfotransferase (GlcNAc6ST). The enzyme sulfates only the hexosamine residues, which is located on a nonreducing end of the KS chain, which indicates that the described modification of the GlcNAc residues takes place during the elongation of the glycan chain [58]. However, after KS polymerization sulfation of the galactose residues occurs, that is catalyzed by a specific galactosylo-6-sulfotransferase [56]. The KS chain may subsequently be modified by fucosylation of sulfated GlcNAc and binding *N*-acetylneuraminic acid by terminal residues of Gal and Gal6S residues [57]. The binding of the *N*-acetylneuraminic acid is likely to finish the biosynthesis of KS and KS II. The terminal residues of KS III are not known [56].

The degradation of keratansulfate PGs initially occurs in the extracellular space, and later, in the lysosomal compartment. In lysosomes, under the influence of acidic hydrolases, like *N*-acetylglucosaminamidase, β-galactosidase, and sulfatases, a gradual removal of subsequent components of the KS chain occurs. Initially, the sulfate group from the last in the KS chain galactose residue is removed, after which the mentioned residue is separated. In the next stage, the hydrolysis covers the sulfate group ester attached to the next GlNAc residue in the degraded chain, which is preceded by a separation of the hexosamine residue from the hydrolyzed KS chain. The reaction is repeated until there is a complete KS chain scission [48].

The keratin sulfate PGs occur in many tissues, while their greatest content is definitely in the cornea, where it occurs in the interstitial matrix as the so-called small leucine-rich PGs—SLRP, i.e., lumican, keratocan, and mimecan [59]. Other KS—fibromodulin and the so-called PRELP occur in cartilage.

Osteoadherin located in the osseous tissue also belongs to the SLRP family [58]. The main cartilage proteoglycan, also in the matrix, bearing KS chains, next to CS chains, is aggrecan [57]. The keratin sulfate glycans occur also on the cell membrane surface and comprise the isoform CD44 and the proteoglycan called SV2, which both are the first described integral KS chains-modified membrane proteins [58]. KSPG occur also in the central nervous system. It seems that currently, after cartilage and cornea, the brain tissue is another site abundant in KSPG. The specific PGs for the nervous tissue are ABAKAN, the mentioned SV2, claustrine, and fosfocan [58].

Keratan sulfate—the least known of all types of GAGs, similar to other glycans, performs also important functions in the body. The macromolecules are key components of the corneal stroma participating in the regulation of the tissue architecture by interactions with fibrous collagen controlling the size and spatial arrangements of the mentioned fibrils indispensable for specific properties of cornea [58]. In cartilage together with collagen type I, KSPG give a tissue special properties of moving significant loads and counteracting the compressive strengths [57]. They take part in the nervous system metabolism performing also a considerable role in the tissue damage repair [58].

HYALURONIC ACID

Hyaluronic acid (hyaluronan) is a nonsulfated, negatively charged biopolymer formed of disaccharide units containing GlcA and *N*-acetylglucosamine [$\rightarrow$4Gl cAβ1$\rightarrow$3GlcNAcβ1$\rightarrow$] [52]. The HA molecules, built of hundreds or even thousands of sugar subunits, form one-of-a-kind glycosaminoglycan, since their disaccharide components are not subject to any modifications, i.e., sulfation or epimerization as it happens in the other types of GAGs [60]. The only variables, referring to the systemic hyaluronan, are the molecule size, which is determined by the polymer length, and the HA concentration and location [60]. Despite its simple structure, hyaluronan interacts with a great number HA-binding proteins, often called hyaladherins, of a varied tissue expression, cellular location, specificity, binding affinity as well as regulatory mechanisms being subject to differentiation, determine the broad spectrum of its action in the body [60]. It occurs in all tissues and body systems of vertebrates as well as in the cell membranes capsules of many pathogenic strains of bacteria [61]. In the organisms of vertebrates, hyaluronan is located in ECM, where it is its mainly carbohydrate component, on the cell surface, as well as intracellularly [61].

HA is degraded in the course of inflammatory processes, tissue destruction, and oncogenesis. HA's micromolecular fragments may induce the expression of proinflammatory genes such as MIP-1α, MIP-1β, KC, RANTES, MCP-1, IL-8, INF-induced protein 10, as well as with a progressive cytokines' expression such as IL-12 or TNF-α, in endothelial, epithelial, and dendritic cells, in fibroblasts, macrophages [62]. Hyaluronan fragments may also induce the biosynthesis of matrix-degrading enzymes such as metalloproteinases, and nitric oxide synthase (iNOS), or inhibitor-1 plasminogen activator [53]. HA oligomers also stimulate cell migration, proliferation and maturation, and manifest proangiogenic activity [54]. On the other hand, the native HA chains with a high molecular mass, inhibit proliferation, migration, and show an antiangiogenic, antiinflammatory, and immunosuppressive activity [62]. The native, biologically inactive with a high molecular mass, negatively charged, unbranched hyaluronan polymers, commonly occurring in the ECM of all animal tissue, play a significant role in maintaining the tissue structural integrity, sequestration of reactive oxygen species, extravascular distribution of plasma proteins, and water-electrolyte homeostasis of the extracellular space [46].

Extraordinary physicochemical properties of hyaluronan, resulting from the combination of its spatial structure, taking the form of a spiral with random coils, are responsible for its strong hygroscopic properties (HA is one of the most hygroscopic molecules occurring in nature, binding water in the amount 1000 times exceeding its mass). Hence, a macromolecule of this GAG takes a huge hydrodynamic volume creating highly viscous and flexible solutions filling the intercellular space and constituting a physiological "lubricant" for articular surfaces or tendon sheaths [55]. HA also participates in controlling skin hydration both in different periods of ontogenesis as well as in pathological conditions, e.g., inflammation processes [55]. Hyaluronan forms half of skin's mass, where this natural polyanion polysaccharide occupies a significant part of its ECM, determining its elasticity, moisture, the desired structure, and moreover providing a means of transport for ions and nutrients [63]. Due to the high molecular mass, considerable sizes of hyaluronan polymers, and hygroscopic properties, this GAG plays a significant role in maintaining a structural and biomechanics tissue integrity, forming at the same time a stroma for cell migration, differentiation, and proliferation [60]. The HA functions are performed by interactions with a huge number of specific proteins, the so-called Hyaladherins [61]. For example, aggrecan together with HA forms well-known, massive, multimolecular proteoglycan aggregates [61]. These expansive complexes play an important role in formation and stabilization of extracellular cartilage matrix generating its viscoelastic properties and giving it the capabilities of supporting significant loads [64]. Complexes of HA with large, aggregating PGs—hyaladherins, occur also in extracartilaginous tissues [63]. For instance, versican was found in skin, while in the nerve tissue hyaluronectin, brevican, and neurocan occur [64]. The mentioned PGs complexes-HA are stabilized also with link proteins, while PGs themselves and the mentioned link proteins are grouped as tissue hyaladherins (hyalectans) [54].

The macromolecules of HA, which is one of the main components of animal tissue ECM, also interact with the glycoproteins present in the matrix, such as fibronectin, laminin, fibril proteins—collagen or elastin, as well as with surface cell receptors "embedded" in the matrix [64]. The latter, also called hyaladherins of the cell surface, comprise a series of specific proteins among which the most well-known hyaladherins are CD44 (one of the markers of breast [cancer] stem cells) and RHAMM [65]. The discovery of both receptors indicated for the first time a direct contribution of hyaluronian in the regulation of cell motility and processes of invasion proliferation [61]. The main cell surface receptor for hyaluronian is CD44. It is a transmembrane glycoprotein occurring on the surface of most cells [65]. A series of isoforms of this hyaladherin reflect the diversity of the functions performed by this "surface" cell receptor [64]. It has been shown that CD44—an integral membrane protein—mediates the stimulation of adhesion, aggregation, migration and cell proliferation, in angiogenesis, in signal transmission on a matrix-cell, and cell-matrix route or in internalization and HA degradation [54]. The second of the mentioned proteins is RHAMM, also known as CD168 receptor, binding HA to a considerably lesser extent, and occurs in many isoforms. Apart from being located on the cell surface, where it is also called the CD168 receptor, it also occurs in cytosol within the cytoskeleton and in the nucleus [54]. This protein mediates in migration and proliferation of normal and cancerous cells. HA also binds with the intercellular adhesion molecule ICAM-1 also called the receptor CD54 and previously as a metabolic receptor for HA, which is responsible mainly for mediating the activation of inflammatory processes [54].

Apart from tissue- and cell surface hyaladherins, in the system there are also hyaladherins located intracellularly [66, 67]. Three specific intracellular receptors for HA, apart from the intracellular isoform of the RHAMM receptor have been well defined. These are CDC37, which accounts for the participation in the cellular cycle regulation and in the regulation of cellular kinases' activity; P-32 receptor (also called HABP-1) responsible for cell signal transduction and most likely in the interactions nucleus-mitochondria; the third protein—IHABP4 receptor is most likely to participate in the intracellular signaling [68]. Hyaladherins mediate the interaction of hyaluronan with the cell surface. This glycosaminoglycan may also be "linked" with the cell surface by a transmembrane interaction with hyaluronan synthases (HASs) [59]. HASs are glycosyltransferases catalyzing the biosynthesis of HA occurring in the internal surface of the plasma membrane of bacterial and eukaryotic cells. These transmembrane enzymatic proteins, including in mammals a group of three enzymes (HAS-1, HAS-2, and HAS-3), each possess two enzymatic components, the first of which catalyzes the transglycosylation of GlcA, while the second *N*-acetyloglucosamine (GlcNAc) out of appropriate nucleotide precursors, i.e., UDP-GlcA and UDP-GlcNAc alternately attach the monosaccharide subunits in question to the reducing end of the growing hyaluronan chain [46]. Growing on the internal, cytoplasmic side of the cell membrane, the HA chain from the side of the nonreducing end is subject to translocation outside the cell membrane, while its synthesis is still continuing [59]. This unusual mechanism of hyaluronan biosynthesis leads to a formation of polymers of "an unlimited" length and therefore with a huge molecular mass, of approximately 10^7 Da. The properties underlie many specific physiological functions in the system [46, 54]. A distinct pattern of biosynthesis compared to the other GAGs causes the biopolymer to form a pericellular capsule, which is at least important for the antioxidative role of the HA function, while creating a pericellular environment with a desired rate of hydration [54, 59]. A distinct chromosomal location of the genes coding each of the three HAS isoenzymes is tissue specific. Their distribution reflects different demand for monosaccharide substrates, different catalytic properties of these isoenzymes, and a varied length of the final product of the HA polymer due to the difference in enzymatic activity [55]. HAS-1 and HAS-3 catalyze the biosynthesis of polymers with a smaller length of the chain, with a smaller molecular mass, compared to HAS-2 products [46]. Hyaluronan molecules with a small molecular mass are more effective in linking intracellular signal transduction with cell surface hyaladherins than those with a bigger mass. The varied expression of hyaluronan syntheses and the regulation of the enzymatic activity have significant consequences in modulating the functioning of a cell [61].

An integral element of HA metabolism is its systemic degradation. The speed of HA catabolism depends on the kind of tissue [61]. Thus, the half-life of this biopolymer in blood equals to 2–5 min for 12 h in skin, from 1 to 3 week in cartilage and 70 days in the eye vitreous body [46, 61]. In the Ha catabolism process, a part of this polymer is subject to in situ degradation, in the place of its biosynthesis and occurrence in tissue, the rest is transported to lymph nodes with lymph and degraded there, while the HA residue not degraded in the mentioned lymphatic structure reaches the general circulation from which it is removed by liver sinusoidal endothelial cells, and to the minimal level also by kidneys [61]. In the course of the local degradation, elimination of HA from tissues occurs via receptors, mainly performed via CD44, endocytosis, and is often preceded by an extracellular degradation catalyzed by extracellular hyaluronidases or activity of reactive oxygen species [61]. In the cell, HA is degraded within lysosomes under the influence of hyaluronidase and exoglycosidase, such as β-glucuronidase or hexosaminidase [69]. A few isoenzymes of hyaluronidase (Hyal) Hyal 1, Hyal 2, Hyal 3 i PH 20 have been described [61]. The HA catabolism occurring in the lymph nodes is facilitated by the LYVE-1 receptor, which is only expressed in the lymphatic endothelial cell surface, while LEC receptor participates in blood cleansing from the circulating hyaluronan. The LEC receptor is located on the surface of endothelial cell of liver sinusoidal vessels [61].

SUMMARY—APPLICATION OF GAGs

A high negative charge density and the capacity for binding various molecules make HA an attractive biomaterial for tissue engineering and drug delivery [66]. A unique, viscoelastic nature of HA together with its biocompatibility and lack of immunogenicity has led to its many clinical applications including supplementation of the joint fluid in inflammation, in eye surgery, and facilitation of healing and regeneration after surgical procedures [66, 70]. In addition, unmodified HA has many important applications in drug delivery, e.g., it is used as an adjuvant for ophthalmic drugs, and enhances drugs and proteins absorption through a mucous membrane [66]. HA was also studied as a drug delivery facilitating agent for different routes of administration including nasal, pulmonary, parenteral, and topical. 3% of diclofenac in 2.5% HA gel (Solaraze) was approved for topical use in solar keratosis, which is the third most frequent skin ailment in the United States. The gel is well tolerated and safe. It is an attractive and cost-efficient alternative method of treatment of solar keratosis [54, 70]. In addition, a randomized clinical study was conducted with a 0.2% HA-based cream (IaLugen) in patient after radiotherapy. The application of the HA cream twice daily significantly decreased the intensity of severe skin reactions compared to placebo. Moreover, due to the application of the cream with hyaluronan, the healing process is more effective [54, 71]. In another multicenter study, patients with venous leg ulcers were treated. They were given an HA gaze rag for 3 weeks compared it with dextranomer paste. Tolerance of both procedures was perfect and although in both groups the appearance and size of ulcers clearly indicated improvement, the process was more effective in the group where hyaluronan was administered [54, 72].

HA is also an attractive building block for new biocompatible and biodegradable palmers applied in drug delivery, tissue engineering, and viscosupplementation [66]. The release of a drug from the HA hydrogel film was assessed in vitro using selected antibacterial and anti-inflammatory drugs. Hence, this new biomaterial could be used in a controlled release of therapeutic agents in the site of injury [66]. The chemical modification allows an adjustment of the physicochemical properties and the residence time of HA in vivo for specific applications. A new HA hydrogel well fit for drug delivery and prepared with the use of a macromolecular crosslinker has been described [66]. The HA derivatives have many possible applications in drug delivery including vestibular gentamicin release [73], pilocarpine delivery into the eye [74], intranasal insulin release [75], and intravaginal calcitonin via HA-ester microspheres [66, 76]. The innovative, fast-acting, and quickly swelling hydrogel film is a naturally, more acceptable biomaterial than unmodified HA hydrogels [66]. Two potential applications of a newly developed HA hydrogel film include wound healing and postoperative adhesions prevention. In order to assess the capacity of the HA hydrogel for a controlled drug release, several antiinflammatory agents were chosen for drug release study. Drug molecules were implemented into the hydrogel film by a polymerization method in situ. The results point out that HA hydrogel film has a perfect potential for a controlled drug release based on the drug hydrophobicity, which enables its action as an ideal device for drug delivery in the injury site [66]. The HA hydrogel film maintains a slow, extended rate of release for hydrophobic antiinflammatory steroids, and therefore it has a potential as a carrier delivering drugs with an extended time release at the injury site [66].

The application of CS/DS and their derivatives in medicine and pharmacy has been limited. CS/DS interact with a broad spectrum of critical molecules such as growth factors, cytokines, chemokines, adhesion molecules by specific saccharide domains present in the chains [67]. Taking into account the variety and physiological importance of the proteins interaction with CS/DS, these GAGs constitute a potential "basis" for the development of new medical and pharmaceutical strategies. A broad biological activity of CS/DS resulting from the mentioned properties might be helpful in designing new CS/DS-based therapies. For instance, making use of the fact that particular CS variants seems to be involved in infections of different microorganisms, which suggests that the CS/DS oligosaccharide sequences specifically interact with the microorganisms may lead to a development of new drugs inhibiting infections [67]. The biological activity of CS/DS chains may include different growth factors. Specific CS/DS chains recruit growth factors and strengthen their activity thanks to which small amounts of CS/DS functional chains may be used for tissue regeneration instead of signal proteins [67]. Apart from that, specific saccharide structures in the CS/DS chains seem to be involved in proliferation and metastases of neoplastic cells. Detection and identification of such CD/DS saccharide sequences constitutes a significant contribution in cancer treatment [67].

The identification of the role of functional domains in the CS/DS structure is necessary for designing and implementing innovative drug dosage forms, medical materials, and bioprostheses for common use [36, 67]. One of the already used medical materials is the matrix containing in its structure beside CS also collagens, keratinocytes, and fibroblasts. It can be applied in adults for regeneration of different organs, in surgical burn treatment, and also in aesthetic medicine [36, 43]. Another biomaterial used in the therapy of tissue injuries is the ECM consisting of CS, DS, HA and also collagens type I, III, IV and fibronectin [36, 77, 78]. The next example of a practical application of DS and HS in pharmacy and medicine is sulodexide, a pleiotropic drug containing 20% of DS and 80% of HS. The formulation is used in, among others, the therapies of slow healing wounds, for instance, as a result of a chronic venous insufficiency [36, 79].

New generations of scaffolds are more and more frequently used as a carrier for delivering therapeutic molecules such as drugs, proteins, and genes, which initiate biological events leading to tissue regeneration [80]. GAGs may be isolated among others from natural source, i.e., animal tissues, and used in biocompatible scaffolds [81]. GAGs favor cell proliferation; hence, they may serve as a collagen scaffold improvement in the tissue bioengineering, e.g., in bone regeneration by promoting angiogenesis [80, 81]. The components consisting gelatin, hyaluronan, and heparin have also demonstrated the efficacy in providing growth factors in mice among others by stimulating the formation of blood vessels [82]. The inclusion of gelatin into the hyaluronan scaffolds has increased the cell proliferation in a dose-dependent manner. Additionally, gelatin and hyaluronan may be mixed by electrospinning in order to create biodegradable sheets of tunable surface voltages, which may be used as drug carrier [82].

REFERENCES

[1] Parenteau-Bareil R, Gauvin R, Berthod F. Collagen-based biomaterials for tissue engineering applications. Materials 2010;3:1863–87.

[2] Canty EG, Kadler KE. Procollagen trafficking, processing and fibrillogenesis. J Cell Sci 2005;118(Pt 7):1341–53, 15788652.

[3] Czubak KA, Żbikowska HM. Struktura, funkcja i znaczenie biomedyczne kolagenów. Ann Acad Med Siles 2014;68(4):245–54.

[4] Ricard-Blum S. The collagen family. Cold Spring Harb Perspect Biol 2011;3(1):a004978, 21421911.

[5] van der Rest M, Garrone R. Collagen family of proteins. FASEB J 1991;5(13):2814–23, 1916105.

[6] Ricard-Blum S, Ruggiero F. The collagen superfamily: from the extracellular matrix to the cell membrane. Pathol Biol (Paris) 2005;53(7):430–42, 16085121.

[7] Abe A, et al. Polymer composites—polyolefin fractionation—polymeric peptidomimetics—collagens. Berlin, Heidelberg: Springer-Verlag; 2018.

[8] Chevallay B, Herbage D. Collagen-based biomaterials as 3D scaffold for cell cultures: applications for tissue engineering and gene therapy. Med Biol Eng Comput 2000;38(2):211–8, 10829416.

[9] Sharma S, Poddar R, Sen P, Andrews J. Effect of vitamin C on collagen biosynthesis and degree of birefringence in polarization sensitive optical coherence tomography (PS-OCT). Afr J Biotechnol 2010;7:2049–54.

[10] Kaku M, Yamauchi M. Mechano-regulation of collagen biosynthesis in periodontal ligament. J Prosthodont Res 2014;58(4):193–207.

[11] Greenspan DS. Biosynthetic processing of collagen molecules. In: Brinckmann J, Notbohm H, Müller PK, editors. Collagen. Topics in Current Chemistry, vol. 247. Berlin, Heidelberg: Springer; 2018.

[12] Aguda AH, Panwar P, Du X, Nguyen NT, Brayer GD, Bromme D. Structural basis of collagen fiber degradation by cathepsin K. Proc Natl Acad Sci U S A 2014;111(49):17474–9, 25422423.

[13] Chattopadhyay S, Raines RT. Collagen-based biomaterials for wound healing. Biopolymers 2014;101(8):821–33, 24633807.

[14] Fields GB. A model for interstitial collagen catabolism by mammalian collagenases. J Theor Biol 1991;153(4):585–602, 1666905.

[15] Weadock KS, Miller EJ, Keuffel EL, Dunn MG. Effect of physical crosslinking methods on collagen-fiber durability in proteolytic solutions. J Biomed Mater Res 1996;32(2):221–6, 8884499.

[16] Harriger MD, Supp AP, Warden GD, Boyce ST. Glutaraldehyde crosslinking of collagen substrates inhibits degradation in skin substitutes grafted to athymic mice. J Biomed Mater Res 1997;35(2):137–45, 9135162.

[17] Wu X, Black L, Santacana-Laffitte G, Patrick Jr. CW. Preparation and assessment of glutaraldehyde-crosslinked collagen-chitosan hydrogels for adipose tissue engineering. J Biomed Mater Res A 2007;81(1):59–65, 17109417.

[18] Weadock KS, Miller EJ, Bellincampi LD, Zawadsky JP, Dunn MG. Physical crosslinking of collagen fibers: comparison of ultraviolet irradiation and dehydrothermal treatment. J Biomed Mater Res 1995;29(11):1373–9, 8582905.

[19] Wedge CI, Rootman DS. Collagen shields: efficacy, safety and comfort in the treatment of human traumatic corneal abrasion and effect on vision in healthy eyes. Can J Ophthalmol 1992;27(6):295–8, 1451017.

[20] Spikes JD, Shen HR, Kopeckova P, Kopecek J. Photodynamic crosslinking of proteins. III. Kinetics of the FMN- and rose bengal-sensitized photooxidation and intermolecular crosslinking of model tyrosine-containing N-(2-hydroxypropyl)methacrylamide copolymers. Photochem Photobiol 1999;70(2):130–7, 10461454.
[21] Ma XH, Noishiki Y, Yamane Y, Iwai Y, Marato D, Matsumoto A. Thermal cross-linking for biologically degradable materials. Preliminary report. ASAIO J 1996;42(5):M866–71, 8945008.
[22] Friess W. Collagen-biomaterial for drug delivery. Eur J Pharm Biopharm 1998;45(2):113–36, 9704909.
[23] Wachol-Drewek Z, Pfeiffer M, Scholl E. Comparative investigation of drug delivery of collagen implants saturated in antibiotic solutions and a sponge containing gentamicin. Biomaterials 1996;17(17):1733–8, 8866036.
[24] Wang MC, Pins GD, Silver FH. Collagen fibres with improved strength for the repair of soft tissue injuries. Biomaterials 1994;15(7):507–12, 7918903.
[25] Friedberg ML, Pleyer U, Mondino BJ. Device drug delivery to the eye. Collagen shields, iontophoresis, and pumps. Ophthalmology 1991;98(5):725–32, 2062508.
[26] Lee CH, Singla A, Lee Y. Biomedical applications of collagen. Int J Pharm 2001;221 (1):1–22.
[27] Reidy JJ, Gebhardt BM, Kaufman HE. The collagen shield. A new vehicle for delivery of cyclosporin A to the eye. Cornea 1990;9(3):196–9, 2373013.
[28] Ruszczak Z, Friess W. Collagen as a carrier for on-site delivery of antibacterial drugs. Adv Drug Deliv Rev 2003;55(12):1679–98, 14623407.
[29] Miller BH, Shavin JS, Cognetta A, Taylor RJ, Salasche S, Korey A, et al. Nonsurgical treatment of basal cell carcinomas with intralesional 5-fluorouracil/epinephrine injectable gel. J Am Acad Dermatol 1997;36(1):72–7, 8996264.
[30] Chen MC, Liang HF, Chiu YL, Chang Y, Wei HJ, Sung HW. A novel drug-eluting stent spray-coated with multi-layers of collagen and sirolimus. J Control Release 2005;108(1):178–89, 16162366.
[31] Mondino BJ. Collagen shields. Am J Ophthalmol 1991;112(5):587–90.
[32] Hoffmann W. In: Dover JS, Carruthers J, Carruthers A, editors. Metody wypełniania tkanek miękkich stosowane w kosmetologii (Seria Dermatologia kosmetyczna). Wrocław: Andrzej Kaszuba/Elsevier Urban & Partner; 2018.
[33] Geiger M. Collagen sponge implants – applications, characteristics and evaluation: Part 1. Pharmaceutical Technology Europe. HighBeam Res 2018.
[34] Johnson Z, Proudfoot AE, Handel TM. Interaction of chemokines and glycosaminoglycans: a new twist in the regulation of chemokine function with opportunities for therapeutic intervention. Cytokine Growth Factor Rev 2005;16(6):625–36, 15990353.
[35] Handel TM, Johnson Z, Crown SE, Lau EK, Proudfoot AE. Regulation of protein function by glycosaminoglycans—as exemplified by chemokines. Annu Rev Biochem 2005;74:385–410, 15952892.
[36] Olczyk PK-VK, Olczyk K. Farmaceutyczny. Przegląd Naukowy. Sci Rev Pharm 2010. Available from http://docplayer.pl/7754836-Farmaceutyczny-przeglad-naukowy-scientific-review-in-pharmacy-www-fpn-sum-edu-pl-lat-cena-24-50-zl.html.
[37] Penc SF, Pomahac B, Winkler T, Dorschner RA, Eriksson E, Herndon M, et al. Dermatan sulfate released after injury is a potent promoter of fibroblast growth factor-2 function. J Biol Chem 1998;273(43):28116–21, 9774430.
[38] Trowbridge JM, Rudisill JA, Ron D, Gallo RL. Dermatan sulfate binds and potentiates activity of keratinocyte growth factor (FGF-7). J Biol Chem 2002;277(45):42815–20, 12215437.

[39] Selleck SB. Proteoglycans and pattern formation: Sugar biochemistry meets developmental genetics. Trends Genet 2000;16(5):206–12, 10782114.
[40] O'Connell N. It's all the buzz. Nurs Stand 2005;20(8):22–4, 16295595.
[41] Volpi N. Therapeutic applications of glycosaminoglycans. Curr Med Chem 2006;13(15):1799–810, 16787222.
[42] Peplow PV. Glycosaminoglycan: a candidate to stimulate the repair of chronic wounds. Thromb Haemost 2005;94(1):4–16, 16113778.
[43] Vynios DH, Papadas Th A, Faraos A, Mastronikolis NS, Goumas P, Tsiganos CP. A solid phase assay for the determination of heparan sulfate and its application to normal and cancerous human cartilage samples. J Immunoass Immunochem 2001;22(4):337–51, 11816802.
[44] Vuillermoz B, Wegrowski Y, Contet-Audonneau JL, Danoux L, Pauly G, Maquart FX. Influence of aging on glycosaminoglycans and small leucine-rich proteoglycans production by skin fibroblasts. Mol Cell Biochem 2005;277(1–2):63–72, 16132716.
[45] Akiyama H., et al. Chondroitin sulfate structure affects its immunological activities on murine splenocytes sensitzed with ovalbumin. Biochem J 2004;382:269–78, 15147241.
[46] Taylor KR, Gallo RL. Glycosaminoglycans and their proteoglycans: host-associated molecular patterns for initiation and modulation of inflammation. FASEB J 2006;20(1):9–22, 16394262.
[47] Kainulainen V, Wang H, Schick C, Bernfield M. Syndecans, heparan sulfate proteoglycans, maintain the proteolytic balance of acute wound fluids. J Biol Chem 1998;273(19):11563–9, 9565572.
[48] Kozma EM, Olczyk K, Wisowski G, Glowacki A, Bobinski R. Alterations in the extracellular matrix proteoglycan profile in Dupuytren's contracture affect the palmar fascia. J Biochem 2005;137(4):463–76, 15858170.
[49] Elizabeth H. Cell biology of extracellular matrix. New York: Plenum Press; 2018.
[50] Nili N, Cheema AN, Giordano FJ, Barolet AW, Babaei S, Hickey R, et al. Decorin inhibition of PDGF-stimulated vascular smooth muscle cell function: potential mechanism for inhibition of intimal hyperplasia after balloon angioplasty. Am J Pathol 2003;163(3):869–78, 12937128.
[51] Gallo RL. Proteoglycans and cutaneous vascular defense and repair. J Investig Dermatol Symp Proc 2000;5(1):55–60, 11147676.
[52] Hileman RE, Fromm JR, Weiler JM, Linhardt RJ. Glycosaminoglycan-protein interactions: definition of consensus sites in glycosaminoglycan binding proteins. Bioessays 1998;20(2):156–67, 9631661.
[53] Scheibner KA, Lutz MA, Boodoo S, Fenton MJ, Powell JD, Horton MR. Hyaluronan fragments act as an endogenous danger signal by engaging TLR2. J Immunol 2006;177(2):1272–81, 16818787.
[54] Weindl G, Schaller M, Schafer-Korting M, Korting HC. Hyaluronic acid in the treatment and prevention of skin diseases: molecular biological, pharmaceutical and clinical aspects. Skin Pharmacol Physiol 2004;17(5):207–13, 15452406.
[55] Chen WY, Abatangelo G. Functions of hyaluronan in wound repair. Wound Repair Regen 1999;7(2):79–89, 10231509.
[56] Głowacki A, Koźma EM, Olczyk K. Biosynteza proteoglikanow keratanosiarczanowych, chondroitynosiarczanowych i dermatanosiarczanowych. Postepy Biochem 2004;50(2):170–81.
[57] Poon CJ, Plaas AH, Keene DR, McQuillan DJ, Last K, Fosang AJ. N-linked keratan sulfate in the aggrecan interglobular domain potentiates aggrecanase activity. J Biol Chem 2005;280(25):23615–21, 15849197.
[58] Funderburgh JL. Keratan sulfate: Structure, biosynthesis, and function. Glycobiology 2000;10(10):951–8, 11030741.

[59] Toole BP. Hyaluronan: from extracellular glue to pericellular cue. Nat Rev Cancer 2004;4(7):528–39, 15229478.
[60] Taylor KR, Trowbridge JM, Rudisill JA, Termeer CC, Simon JC, Gallo RL. Hyaluronan fragments stimulate endothelial recognition of injury through TLR4. J Biol Chem 2004;279(17):17079–84, 14764599.
[61] Tammi MI, Day AJ, Turley EA. Hyaluronan and homeostasis: a balancing act. J Biol Chem 2002;277(7):4581–4, 11717316.
[62] Slevin M, Kumar S, Gaffney J. Angiogenic oligosaccharides of hyaluronan induce multiple signaling pathways affecting vascular endothelial cell mitogenic and wound healing responses. J Biol Chem 2002;277(43):41046–59, 12194965.
[63] Oishi Y, Kato H, Noguchi T. Dietary protein as a potent regulator of the hyaluronan synthase gene in rat skin(food & nutrition science). Biosci Biotechnol Biochem 2003;67(4):736–42.
[64] Turino GM, Cantor JO. Hyaluronan in respiratory injury and repair. Am J Respir Crit Care Med 2003;167(9):1169–75, 12714341.
[65] Kikuchi S, Griffin CT, Wang SS, Bissell DM. Role of CD44 in epithelial wound repair: migration of rat hepatic stellate cells utilizes hyaluronic acid and CD44v6. J Biol Chem 2005;280(15):15398–404, 15691832.
[66] Luo Y, Kirker KR, Prestwich GD. Cross-linked hyaluronic acid hydrogel films: new biomaterials for drug delivery. J Control Release 2000;69(1):169–84, 11018555.
[67] Yamada S, Sugahara K. Potential therapeutic application of chondroitin sulfate/dermatan sulfate. Curr Drug Discov Technol 2008;5(4):289–301, 19075609.
[68] Hascall VC, Majors AK, De La Motte CA, Evanko SP, Wang A, Drazba JA, et al. Intracellular hyaluronan: a new frontier for inflammation? Biochim Biophys Acta 2004;1673(1–2):3–12, 15238245.
[69] Gushulak L, Hemming R, Martin D, Seyrantepe V, Pshezhetsky A, Triggs-Raine B. Hyaluronidase 1 and beta-hexosaminidase have redundant functions in hyaluronan and chondroitin sulfate degradation. J Biol Chem 2012;287(20):16689–97, 22451654.
[70] Brown MB, Jones SA. Hyaluronic acid: a unique topical vehicle for the localized delivery of drugs to the skin. J Eur Acad Dermatol Venereol 2005;19(3):308–18, 15857456.
[71] Liguori V, Guillemin C, Pesce GF, Mirimanoff RO, Bernier J. Double-blind, randomized clinical study comparing hyaluronic acid cream to placebo in patients treated with radiotherapy. Radiother Oncol 1997;42(2):155–61, 9106924.
[72] Ortonne J. A controlled study of the activity of hyaluronic acid in the treatment of venous leg ulcers. J Dermatolog Treat 1996;7:75–81. https://doi.org/10.3109/09546639609089533.
[73] Kelly RM, Meyer JD, Matsuura JE, Shefter E, Hart MJ, Malone DJ, et al. In vitro release kinetics of gentamycin from a sodium hyaluronate gel delivery system suitable for the treatment of peripheral vestibular disease. Drug Dev Ind Pharm 1999;25(1):15–20, 10028414.
[74] Saettone MF, Giannaccini B, Chetoni P, Torracca MT, Monti D. Evaluation of high- and low-molecular-weight fractions of sodium hyaluronate and an ionic complex as adjuvants for topical ophthalmic vehicles containing pilocarpine. Int J Pharm 1991;72(2):131–9.
[75] Illum L, Farraj NF, Fisher AN, Gill I, Miglietta M, Benedetti LM. Hyaluronic acid ester microspheres as a nasal delivery system for insulin. J Control Release 1994;29(1):133–41.
[76] Rochira M, Miglietta MR, Richardson JL, Ferrari L, Beccaro M, Benedetti L. Novel vaginal delivery systems for calcitonin: II. Preparation and characterization of HYAFF® microspheres containing calcitonin. Int J Pharm 1996;144(1):19–26.
[77] Harley BA, Leung JH, Silva EC, Gibson LJ. Mechanical characterization of collagen-glycosaminoglycan scaffolds. Acta Biomater 2007;3(4):463–74, 17349829.

[78] Wounds. Does bilayered extracellular matrix technology hasten wound healing in venous stasis ulcers? A retrospective study. Available from http://www.woundsresearch.com/article/3755; 2018.

[79] Jacek Borawski MM. New pharmacological effects of heparin and sulodexide. Postępy Nauk Medycznych 2009.

[80] Quinlan E, Partap S, Azevedo MM, Jell G, Stevens MM, O'Brien FJ. Hypoxia-mimicking bioactive glass/collagen glycosaminoglycan composite scaffolds to enhance angiogenesis and bone repair. Biomaterials 2015;52:358–66, 25818442.

[81] Kaczmarek B, Sionkowska A, Osyczka AM. Collagen-based scaffolds enriched with glycosaminoglycans isolated from skin of *Salmo salar* fish. Polym Test 2017;62:132–6.

[82] Santoro M, Tatara AM, Mikos AG. Gelatin carriers for drug and cell delivery in tissue engineering. J Control Release 2014;190:210–8, 24746627.

Chapter 12

Examples of Successful Biomaterial-Based Artificial Tissues—Artificial Corneas

Bartosz Sikora[*,†], Mehrdad Rafat[‡,§], Marek J. Łos[¶,‖]

*Department of Molecular Biology, School of Pharmacy with the Division of Laboratory Medicine in Sosnowiec, Medical University of Silesia, Katowice, Poland

†Department of Cytophysiology, Chair of Histology and Embryology, School of Medicine in Katowice, Medical University of Silesia, Katowice, Poland

‡Department of Biomedical Engineering, Linköping University, Linköping, Sweden

§LinkoCare Life Sciences AB (Ltd), Linköping, Sweden

¶Department of Pathology, Pomeranian Medical University, Szczecin, Poland

‖Center for Molecular Biophysics, UPR4301 CNRS CS80054, Orleans Cedex 2, France

INTRODUCTION

The first decades of the 21st century are marked by rapid development of technologies that advance regenerative medicine. Development of induced-pluripotent stem cell (iPS) technology in 2006 opened the opportunity for clinical applications of stem-cell treatment for broad range of diseases. The reprogramming technology offers new tools that can be used in tissue engineering and whole organ reconstruction [1–4]. Combining the biology with biomaterials resulted in implementation of the stem-cell-based devices in ophthalmology especially in cornea reconstruction [4]. Keratoplasty, or corneal transplantation, offers a leading example that artificial organs/tissues could be produced in vitro and safely implanted to patients [5].

SURGERY PROCEEDINGS

Keratoplasty is a broad term that includes all surgery proceedings executed on the cornea. There are numerous distinct methods of keratoplasty. We should distinguish two major types of the cornea surgery: full-thickness keratoplasty also named penetrating keratoplasty (PK or PKP) and lamellar keratoplasty (LK). The main difference is that the surgery does not penetrate the cornea in its full thickness

Stem Cells and Biomaterials for Regenerative Medicine. https://doi.org/10.1016/B978-0-12-812258-7.00012-5

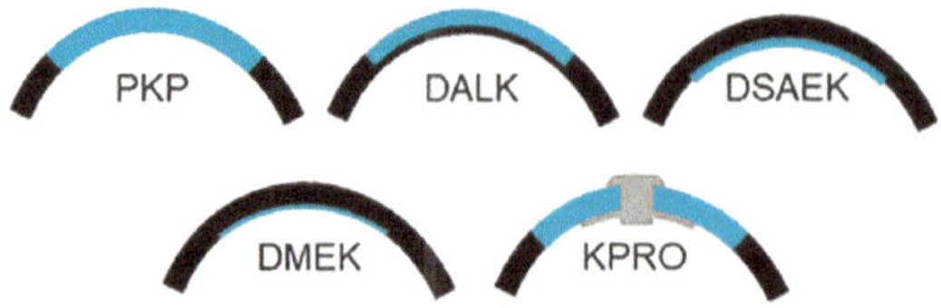

FIG. 12.1 Various modern keratoplasty techniques, including penetrating keratoplasty (PK), deep anterior lamellar keratoplasty (DALK), Descemet stripping automated endothelial keratoplasty (DSAEK), Descemet membrane endothelial keratoplasty (DMEK), and Boston Type I Keratoprosthesis (KPRO). The *blue* marked region is a transplanted part. *(Based on Donaghy CL, Vislisel JM, Greiner MA. An introduction to corneal transplantation; 2015.)*

but includes only selected layers [6, 7]. The first reported PK has been performed in 1905 by Eduard Konrad Zirm on patient with bilateral alkali burns [7]. Since that time, keratoplasty has been improved and has evolved distinctly especially LK [8]. Many different techniques of lamellar keratoplasty have since been developed.

Deep Anterior Lamellar Keratoplasty (DALK) refers to the surgery involving all anterior cornea layers except parts of posterior stroma and Decemet's membrane (DM) [9]. In contrast, DMEK means Descemet Membrane Endothelial Keratoplasty and refers to surgery of DM without all anterior layers. There are also modifications of selected techniques that are usually associated with the tools used and the margin of tissue replaced by keratoplasty. For example, DSAEK stands for Decemet's Stripping Automated Endothelial Keratoplasty [10]. Please see Fig. 12.1 for schematic explanations of various keratinoplasty techniques.

ARTIFICIAL CORNEA CONCEPT

The concept of artificial cornea transplantation is over 200 years old. The first documented data describing this concept are assigned to Guillaume Pellier De Quengsy who was a French ophthalmologist renowned at that time for cataract surgery. In his concept, he described the surgery covering transplantation of very clear glass with similar dimensions to the replaced cornea [11, 12]. Pellier De Quengsy's thought was to replace the disordered cornea with a piece of glass combined with material, which would be appropriate for implantation to peripheral tissues like stroma. The French ophthalmologist proposed a silver ring at the edge of glass as a barrier similar to corneal limbus that would be appropriate to grow into the groove made for adaptation of this implant with stroma [11]. The evidence is lacking that could definitely confirm that Guillaume Pellier De Quengsy has ever performed any surgery similar to what he described. Notwithstanding, his monograph containing precisely described prosthokeratoplasty began the development of this branch in ophthalmology.

KPRO—KERATOPROSTHESIS

Pellier de Quengsy's idea brought new direction to create biocompatible and applicable prosthesis to replace the native damaged corneas in patients with corneal blindness in whom the standard donor keratoplasty does not work. Over the years, Pellier de Quengsy's concept has evolved and resulted in the invention of multiple devices for the treatment of corneal blindness [13]. KPro is a short and international term, which primarily refers to all keratoprosthesis (KPros). Nowadays, KPro defines both the surgery and the manufactured devices for corneal replacement therapies. To be more precise, the short term KPro refers to prosthokeratoplasty consisting of the replacement of damaged cornea with keratoprosthesis (the device) made of synthetic materials, which should be biocompatible with human tissue [14]. In contrast to conventional keratoplasty such as PK, DALK, or DMEK, KPro is an alternative method as a last resort for patients for whom the standard procedure cannot be applied, and the donor tissue cannot be transplanted [15, 16].

Several KPros have been developed in different countries over the past century. We can distinguish three major devices that are the treatment of choice in selected diseases [13].

The most popular device, which is routinely applied, is Boston KPro type 1. It had been invented in early 1950s, and it is based on polymethylmethacrylate (PMMA) material. Subsequently, in the second half of 20th century, several devices for KPro applications have been invented, which were tested in humans, with various degrees of success [13, 15, 17, 18].

The Boston type 1 keratoprosthesis, previously named the Dohlman-Doane device, consists of three elements: the front part, back plate, and metal locking ring made from titanium (Fig. 12.2). The front part of Boston KPro is made from PMMA and is a convex exterior surface with the optics in the center. Those two parts are combined with a locking ring that locks the back plate onto the center optic. The back plate previously was made with PMMA but nowadays, newer versions have been developed that have titanium back plate. Boston

FIG. 12.2 Boston KPro type 1. Left—titanium backplate, right—PMMA backplate, bottom—combined Boston Kpro [19].

KPro with titanium backplate does not require a locking ring. It is formed as an 8.5-mm-diameter disk with a large aperture in the center and multiple peripheral holes with smaller diameter [20]. The backplates role is to keep in place the whole device, hence, improving its retention. Holes are made to improve the nutrition transport through the peripheral and maintaining proper hydration. The Boston KPro has been approved by US Food and Drug Administration (FDA). Nonetheless, this device has limitations. In Boston type 1 KPro, there is still a need for donor tissue. During the graft preparation, donors' corneal button is sandwiched with both parts and then transplanted to the host. Donor tissue here plays the role of a carrier for the whole construct. Boston type 1 KPro is implemented to people with high risk of corneal grafting, in cases with corneal opacities and extensive vascularization or repeat graft failures [13].

The type 2 version of Boston KPro has also been approved by the FDA. The main difference is that the central optics (the stem) is longer than that in the type 1. This method gives the chance to restore vision to patients with severe dry eye diseases, where the type 1 cannot be applied. This type of KPro is less frequently applied as physicians favor the second most popular method—namely, OOKP [21, 22].

OOKP—OSTEOODONTOKERATOPROSTHESIS

This surgery is based on the use of tooth root as a carrier for optical cylinder whose role is to restore vision in severe eye damage (Fig. 12.3). This procedure is being improved since 1960s and is the second most popular keratoprosthesis in the world [24]. The technique is applied to people for whom the traditional corneal transplant from cadaver is not possible because of the size of damage and has similar indications for use as in the Boston type 2 KPro, such as severe dry eyes, Steven-Johnson syndrome (SJS), mucous membrane pemphigoid (MMP), and severe ocular burns that lead to limbal stem cells deficiency [25].

In this procedure, the tooth root and alveolar bone of the patient is used for transplantation as a carrier for optical cylinder [24]. The tooth root plays a role of stabilizer for optical cylinder made from PMMA that goes though the tooth. Osteoodontokeratoprosthesis is a time-consuming procedure divided into multiple steps and requires both oral and eye surgeries. At the first step, the eye is prepared for receiving the osteo-odonto prosthesis. This procedure is essential for making the eye ready for obtaining a graft. At this step, the part of mucosa is transplanted into the eye and attached to the sclera to cover the eye under the eyelid. At further step, after several weeks, the tooth, tooth root and a surrounding bone are extracted. The place is filled with a free bone usually transplanted from the iliac crest. Then the prosthesis is prepared. Preparation of OOKP is a complex procedure, where the tooth root is grinded properly to obtain a desired shape and drilled at the center for making the hole for optical cylinder which is glued in. At the end of prosthesis forming, the tooth crown is grinded out. OOKP prepared in this way then is grafted under the eyelid (usually the lower eyelid)

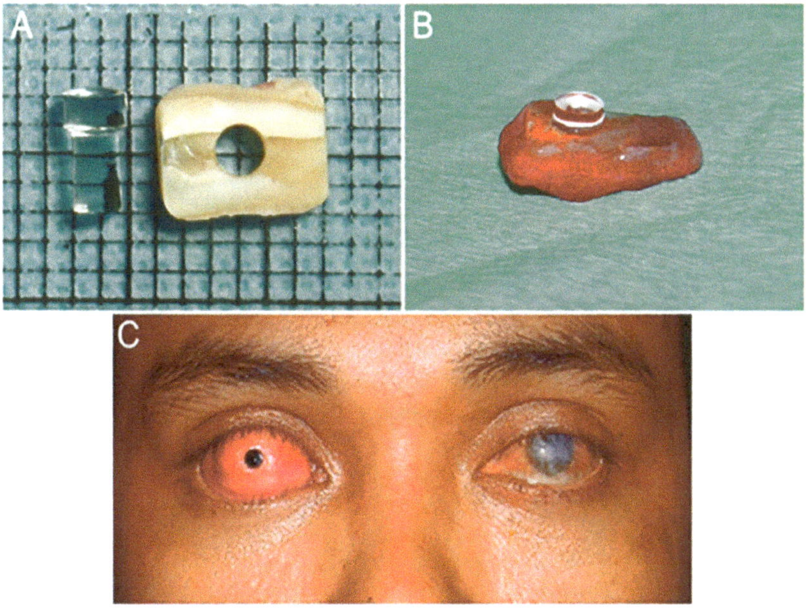

FIG. 12.3 (A) retrieved tooth root grinded and drilled for the cylinder placement, (B) prosthesis with the optical cylinder after fibrosis, (C) healed eye with OOKP [23].

for next 3 months. Since that time, the prosthesis is being fibrosed. When the prosthesis is fully fibrosed, it is extracted again and prepared for proper grafting. The final step covers opening the healed mucosa from the cornea, corneal trephination, complete iridectomy, lens extraction, anterior vitrectomy, prosthesis implantation with fixation to the sclera, globe tonization and finally closing of mucosa and preparing the hole for the optics [24, 26]. For increasing the chance of success of the procedure, this method requires the good conditions of dentition, which in some cases is an issue, especially in elderly.

OKP—OSTEOKERATOPROSTHESIS

Since the OOKP could only be performed by people with healthy dentition, other techniques have been developed for those that do not qualify for OOKP. One such technique involves the KPro preparation based on bones, i.e., tibia [16, 27]. The procedure is similar to osteo-odonto KPro. In this method, the piece of tibia is taken for making the prosthesis. To do this, an incision of skin on the leg in its superior one-third level is made to expose the surface of tibial bone. A round disc of tibia is obtained by rotating trephine drill. The disc has 10 mm in diameter and 3 mm in thickness. After that a piece of bone is grinded and the hole is drilled for PMMA cylinder. Further procedures are the same as in OOKP [16]. In addition to Boston KPros and osteoodontokeratoprosthesis, a third, less popular procedure has been developed, and it is based on the AlphaCor device.

ALPHACOR

This keratoprosthesis differs from both previously mentioned. The main difference is that the whole construct is made from one piece (Fig. 12.4). AlphaCor implant is made from flexible poly-2-hydroxyethyl methacrylate (PHEMA) material, and it is divided into two regions. At the central region, the implant is clear (transparent) surrounded by an opaque skirt. The junction between two phases is bonded on molecular level by polymers interpenetration across a junction, which is known as the interpenetrating polymer network (IPN) [5, 29]. The opaque skirt is porous with high water content. This is designed to improve host cell infiltration and anchoring the prosthesis into place. The AlphaCor device has the approval of Food and Drug administration since 2002. It is designed for people with severe corneal damage such as scarring, vascularization, diseases that are ineligible for standard corneal transplantation, but in contrast the device should be only applied to people with good health of eyelid, proper tear film, and absence of inflammation [29]. As previous KPros, AlphaCor is considered for implantation in multiple-graft failure cases. The procedure is performed in two stages. At the first stage, the keratoprosthesis is placed into the lamellar pocket.

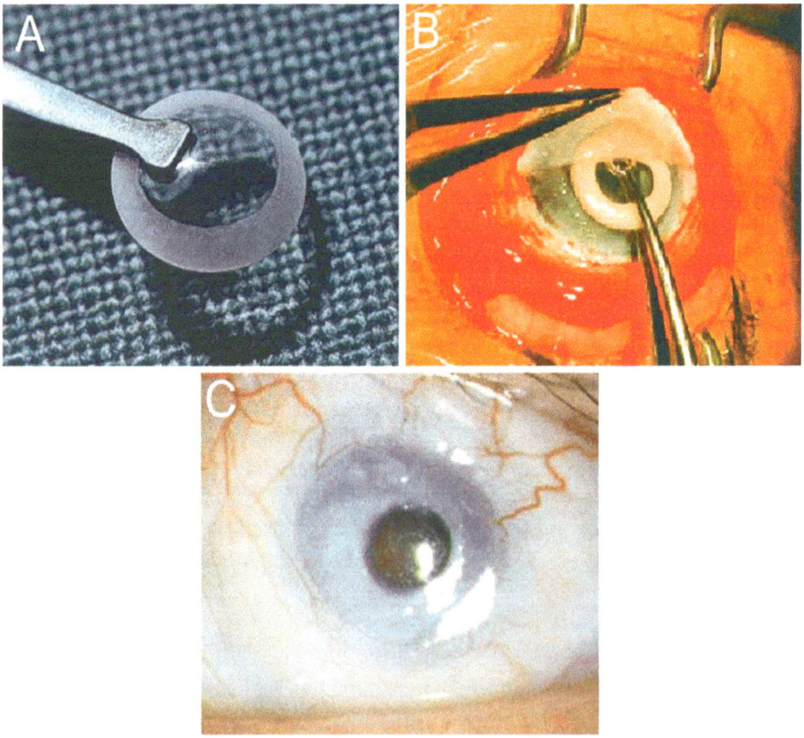

FIG. 12.4 (A) AlphaCor device, (B) implantation of the device into the lamellar pocket, (C) healed eye 24 months after surgery [28].

During this stage, the posterior to the optic corneal tissue is removed and the AlphaCor is left in the pocket for several weeks. At the second stage surgery, when the device is properly nested, the anterior to the optic tissue is removed. At this time the porous skirt should remain enclosed by the stroma, which is possible because of the porous body of the skirt and cellular colonization [30]. The device has similar dimensions to natural donor graft and permits for close to natural light refraction and does not disrupt the intraocular examination.

INDICATIONS FOR ARTIFICIAL CORNEA TRANSPLANTATION

The artificial cornea transplantation concept is older than the development of standard chirurgical methods. However, over last century, the penetrating keratoplasty has been developed and subsequently improved. Besides the penetrating keratoplasty and donor tissue transplantation, relatively less invasive techniques were routinely applied for restoring vision. Lamellar keratoplasty, part corneal tissue transplantation, and several other corneal surgeries are routinely applied therapies for corneal diseases. Notwithstanding, there are multiple indications for the use of artificial devices, which, under certain conditions, gives more promising outcome than standard therapies.

Before choosing the type of therapy, the patient with corneal disease causing blindness should be examined and assessed for potential risks for penetrating keratoplasty. PK is the first considered treatment for corneal blindness. All patients with low risk for PK should be considered for this procedure. In case of failure, the donor graft should be repeated. Multiple graft failure is the major indication for applying Boston KPro type 1. On the other hand, patients with high risk for penetrating keratoplasty should be assessed for the eye surface hydration. If the eye surface is wet, then the KPro type 1 is mainly recommended artificial cornea device. In the case of dry surface of the eye, the health of eye lid gives the direction for consideration the other artificial corneas. Good eyelid health and its serried form are prerequisites for the Boston KPro type 2 or OOKP, which should be considered for transplantation. Under other conditions, when the eye lid is insufficient for complete tarsorrhaphy—the surgery involving eyelid closing by holding it together with sutures [31], then the only OOKP is indicated and given under consideration. Of course, all other devices (e.g., AplhaCor) that could be applied under such indications as mentioned should be considered for transplantation if only it could give more reliable outcomes [5, 13, 21, 22, 29].

DIFFERENCES BETWEEN ARTIFICIAL CORNEAS AND BIOENGINEERED CORNEAS

Standard keratoprosthesis devices differ by the materials and technologies used for their production. It must be underlined that terms artificial corneas and bioengineered corneas should refer to different devices. Standard artificial corneas such as Boston KPros devices are completely synthetic. Those devices consist of solid materials, which generally fill the gap after diseased corneal removal and supports donor corneal graft to properly keep it in place. In this kind of device, its bioretention and biodegradation are not the expected outcomes. The materials used in production of Boston KPros are not biodegradable and its form should be as much permanent as possible for good clinical outcomes after transplantation.

In contrast to artificial corneas, we should distinguish devices, which are designed for cell migration and controlled biodegradation. Despite of osteoodontokeratoprosthesis and similar to them osteokeratoprosthesis, which are the devices combined of host tissues and optical cylinder, there are new inventions, which use the biological potential of natural hosts stem cells, whose main purpose is to support those cells and function as a scaffold. Bioengineered corneas based on hydrogels are the future of keratoplasty, allowing for less invasive surgeries and eliminating the need for donor tissues [32].

HOLOCLAR—THE BIOENGINEERED CORNEAL EPITHELIUM

The newest and the only approved medicinal product in Europe for an ophthalmic indication, based on autologous stem cells, is Holoclar [33]. This device is a bioengineered product developed to make an alternative source for corneal limbal stem cells and corneal epithelium. Corneal cell therapy with limbal stem cells has been described for the first time by Kenyon and Tseng in 1989 [34, 35]. Kenyon and Tseng describe the surgery for cure of limbal stem cells deficiency (LSCD). When the LSCD occurs, the barrier between the eye tissues vanishes that makes serious and painful eye diseases associated with ingrown of peripheral tissues onto the cornea. Since 1989, there are few reports about significant disadvantages of described surgery such as large amount of tissue containing limbal stem cells needed, which is a significant risk for donor eyes that could subsequently become limbal stem cell deficient. Another disadvantage is that the use of allogenic tissues that creates a need for potent immune suppression, which also provides the risk of infections and neoplasia. Since its introduction, the Tseng method has been modified and improved. The major modification was to reduce the size of limbus tissue that needs to be excised, subsequent expansion the limbal cell ex vivo, and then transplantation them into the recipient eye. This manipulation eliminates the need for allografting and gives an opportunity for autologous cell transplant, which also eliminates the need for immune suppression [34].

Such procedure was performed for the first time in 1997 on two patients who for the first time received human limbocorneal epithelium reconstructed in vitro [4]. Since 2004, the procedure became a routine treatment of patients with LSCD in Italy and accepted in India in 2008. In 2015, Holoclar finally has been authorized by European Commission for the use in all the partner countries [33].

The procedure covers several steps. First, the small piece, approximately 2×2 mm of healthy corneal limbus, is removed from patients' eye. The tissue is transported in optimized medium with an addition of antibiotics within 24 h to laboratory of HoloStem company in Italy. Cells then are cultured for several days and cryopreserved for further use in temporary cell bank. For preparing the ready-for-transplant graft, cells are recultured in special, optimized medium but now without any addition of antibiotics. At this time, cells are seeded on fibrin hydrogel biomaterial and cultured upon total coverage. Fibrin scaffold when overgrown with cells, since that time, is Holoclar medicinal product, ready for transplantation. Bioengineered tissue is then transported to the clinic for transplantation [4, 33].

LINKCOR® BIOENGINEERED CORNEA

New ideas for producing bioengineered corneas are still evolving. One of the most promising bioengineered devices is a corneal implant called LinkCor developed by LinkoCare Life Sciences Ltd. in Sweden. This product closely mimics the human corneal stroma and is made of the same material that comprises most of the human cornea, i.e., collagen type I. The preclinical studies and clinical outcomes of the phase I clinical trials for this device are very promising [36, 37]. LinkCor® at the moment, is going through the regulatory process for approval and marketing authorizations in Europe and elsewhere. Collagen is the natural component of corneal extracellular matrix (ECM) and is widely explored and confirmed as the potential material of choice in tissue engineering. LinkCor is one of the first collagen-based devices that is biocompatible, mechanically robust, and implantable that allows cell growth and tissue regeneration. This product's features are comparable to those of human cornea (Fig. 12.5) and bring hope for the use as an alternative to donor grafts [38].

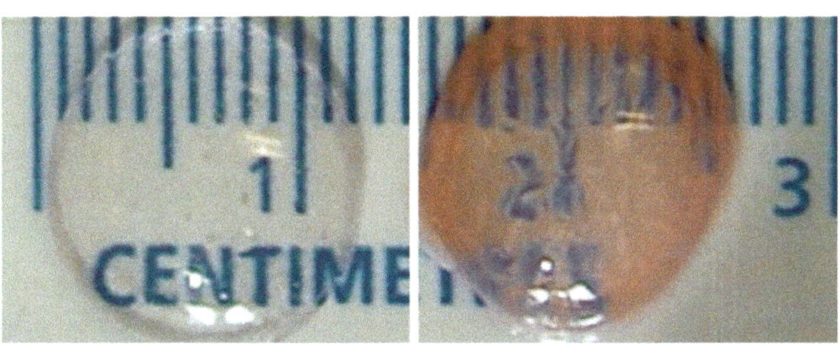

FIG. 12.5 Comparison of LinkCor® Bioengineered Cornea—left, with donor human cornea—right [32].

CLOSING REMARKS

Nowadays, a few successful biomaterial-based artificial corneas are becoming available as alternatives for human (cadaver) donor corneas. There are three kind of artificial corneas routinely applied in the treatment of corneal blindness. More and more artificial equivalents of natural corneal grafts are developed. Such medical devices (medicinal products) face rigorous regulations so that they are sufficiently safe for implementation in standard therapies in ophthalmology. The testing procedures and regulatory approval process extend the time for adapting new bioengineered corneas.

The regenerative medicine as a multidisciplinary branch of science combines stem cells research, physiology, biomaterial sciences, and even (electro)mechanics. Bioengineered corneas create great environment for the growth of stem cells that can restore damaged tissues. Since 2012, when Yamanaka and Gurdon received the Nobel Prize in Physiology or Medicine for the discovery that mature cells could be reprogrammed to become pluripotent or iPSC (induced pluripotent stem cells). Subsequently, numerous protocols describing the differentiation of iPSC to corneal epithelium have been described. Such technologies bring hope for the use of autologous iPSC in cornea treatment [39–41]. Besides iPSC as a source of autologous limbal cells protocols have been developed to directly change fibroblasts into limbal cells in the process of transdifferentiation [39]. Transdifferentiation appears to be safer (no iPSC stage) and faster (only 7–10days) than protocols involving cells differentiation into limbal cells from iPSC (several weeks). Hopefully, combining the achievements in cell reprogramming and tissue engineering will allow for bioengineering of devices totally independent from the need for donor tissues. This would simplify the multistage surgeries and become more comfortable for patients and ophthalmologists.

REFERENCES

[1] Asai A, Aihara E, Watson C, Mourya R, Mizuochi T, Shivakumar P, Phelan K, Mayhew C, Helmrath M, Takebe T, et al. Paracrine signals regulate human liver organoid maturation from induced pluripotent stem cells. Development 2017;144:1056–64.

[2] Howard D, Buttery LD, Shakesheff KM, Roberts SJ. Tissue engineering: strategies, stem cells and scaffolds. J Anat 2008;213:66–72.

[3] Lim ML, Jungebluth P, Ajalloueian F, Friedrich LH, Gilevich I, Grinnemo KH, Gubareva E, Haag JC, Lemon G, Sjoqvist S, et al. Whole organ and tissue reconstruction in thoracic regenerative surgery. Mayo Clin Proc 2013;88:1151–66.

[4] Pellegrini G, Lambiase A, Macaluso C, Pocobelli A, Deng S, Cavallini GM, Esteki R, Rama P. From discovery to approval of an advanced therapy medicinal product-containing stem cells, in the EU. Regen Med 2016;11:407–20.

[5] Akpek EK, Alkharashi M, Hwang FS, Ng SM, Lindsley K. Artificial corneas versus donor corneas for repeat corneal transplants. Cochrane Database Syst Rev 2014;CD009561.

[6] Khattak A, Nakhli FR, Al-Arfaj KM, Cheema AA. Comparison of outcomes and complications of deep anterior lamellar keratoplasty and penetrating keratoplasty performed in a large group of patients with keratoconus. Int Ophthalmol 2018;38:985–992.

[7] Price MO, Gupta P, Lass J, Price FW. EK (DLEK, DSEK, DMEK): new Frontier in cornea surgery. Annu Rev Vis Sci 2017;3:69–90.

[8] Donaghy CL, Vislisel JM, Greiner MA. An introduction to corneal transplantation. EyeRounds [service online] 2015 May 21 [cited 2017 Sep 30]. Available from: http://EyeRounds.org/tutorials/cornea-transplant-intro/

[9] Parthasarathy A, Por YM, Tan DTH. Use of a "small-bubble technique" to increase the success of Anwar's "big-bubble technique" for deep lamellar keratoplasty with complete baring of Descemet's membrane. Br J Ophthalmol 2007;91:1369–73.

[10] Quigley C, McElnea E, Fahy G. Trends in corneal transplant surgery in Ireland: indications and outcomes of corneal transplant surgery and intraocular lens opacification following Descemet's stripping automated endothelial keratoplasty. Irish J Med Sci 2018;187: 231–236.

[11] Chirila TV, Hicks CR. The origins of the artificial cornea: Pellier de Quengsy and his contribution to the modern concept of keratoprosthesis. Gesnerus 1999;56:96–106.

[12] Pellier De Quengsy G. Précis ou cours d'opérations sur la chirurgie des yeux: puisé dans le sein de la pratique, & enrichi de figures en taille-douce. vol. 1. Paris: Didot; 1789.

[13] Avadhanam VS, Smith HE, Liu C. Keratoprostheses for corneal blindness: a review of contemporary devices. Clin Ophthalmol (Auckland, NZ) 2015;9:697–720.

[14] Aquavella JV, Qian Y, McCormick GJ, Palakuru JR. Keratoprosthesis: current techniques. Cornea 2006;25:656–62.

[15] Ang M, Man R, Fenwick E, Lamoureux E, Wilkins M. Impact of type I Boston keratoprosthesis implantation on vision-related quality of life. Br J Ophthalmol 2018;102:878–881.

[16] Charoenrook V, Michael R, de la Paz MF, Ding A, Barraquer RI, Temprano J. Osteokeratoprosthesis using Tibial bone: surgical technique and outcomes. Ocul Surf 2016;14:495–506.

[17] Fry M, Aravena C, Yu F, Kattan J, Aldave AJ. Long-term outcomes of the Boston type I keratoprosthesis in eyes with previous herpes simplex virus keratitis. Br J Ophthalmol 2018;102:48–53.

[18] Muzychuk AK, Robert M-C, Dao S, Harissi-Dagher M. Boston keratoprosthesis type 1. Ophthalmology 2017;124:20–6.

[19] Todani A, Ciolino JB, Ament JD, Colby KA, Pineda R, Belin MW, Aquavella JV, Chodosh J, Dohlman CH. Titanium back plate for a PMMA keratoprosthesis: clinical outcomes. Graefes Arch Clin Exp Ophthalmol 2011;249:1515–8.

[20] Noel CW, Isenberg J, Goldich Y, Conlon R, Teichman J, Rubinger DA, Weinberg TAI, Slomovic J, Chan CC, Rootman DS, et al. Type 1 Boston keratoprosthesis: outcomes at two Canadian centres. Can J Ophthalmol 2016;51:76–82.

[21] Avadhanam VS, Liu CSC. A brief review of Boston type-1 and osteo-odonto keratoprostheses. Br J Ophthalmol 2015;99:878–87.

[22] Zarei-Ghanavati M, Avadhanam V, Vasquez Perez A, Liu C. The osteo-odonto-keratoprosthesis. Curr Opin Ophthalmol 2017;28:397–402.

[23] Tay ABG, Tan DTH, Lye KW, Theng J, Parthasarathy A, Por YM. Osteo-odonto-keratoprosthesis surgery: a combined ocular-oral procedure for ocular blindness. Int J Oral Maxillofac Surg 2007;36:807–13.

[24] Liu C, Paul B, Tandon R, Lee E, Fong K, Mavrikakis I, Herold J, Thorp S, Brittain P, Francis I, et al. The osteo-odonto-keratoprosthesis (OOKP). Semin Ophthalmol 2005;20: 113–28.

[25] Moussa S, Reitsamer H, Ruckhofer J, Grabner G. The ocular surface and how it can influence the outcomes of keratoprosthesis. Curr Ophthalmol Rep 2016;4:220–5.

[26] Hille K, Grabner G, Liu C, Colliardo P, Falcinelli G, Taloni M, Falcinelli G. Standards for modified osteoodontokeratoprosthesis (OOKP) surgery according to Strampelli and Falcinelli: the Rome-Vienna protocol. Cornea 2005;24:895–908.

[27] De La Paz MF, De Toledo JÁ, Charoenrook V, Sel S, Temprano J, Barraquer RI, Michael R. Impact of clinical factors on the long-term functional and anatomic outcomes of osteo-odonto-keratoprosthesis and tibial bone keratoprosthesis. Am J Ophthalmol 2011;151:829–39. [e821].

[28] Hicks CR, Crawford GJ, Lou X, Tan DT, Snibson GR, Sutton G, Downie N, Werner L, Chirila TV, Constable IJ. Corneal replacement using a synthetic hydrogel cornea, AlphaCor (TM): device, preliminary outcomes and complications. Eye 2003;17:385.

[29] Jirásková N, Rozsival P, Burova M, Kalfertova M. AlphaCor artificial cornea: clinical outcome. Eye 2011;25:1138–46.

[30] Hicks CR, Crawford GJ, Dart JKG, Grabner G, Holland EJ, Stulting RD, Tan DT, Bulsara M. AlphaCor: clinical outcomes. Cornea 2006;25:1034–42.

[31] Rajak S, Rajak J, Selva D. Performing a tarsorrhaphy. Commun Eye Health 2015;28:10–1.

[32] Rafat MA, Hackett JM, Fagerholm P, Griffith M. Artificial Cornea. In: Dartt DA, editor-in-chief. Encyclopedia of the Eye. 1st ed. Academic Press: London; 2010. p. 128–134.

[33] Milazzo G, Ardigò D, Toschi M, Matuska S, Rama P, De Luca M, Pellegrini G. Holoclar®: first of its kind in more ways than one. Cell Gene Ther Insights 2016;2:183–97.

[34] Baylis O, Figueiredo F, Henein C, Lako M, Ahmad S. 13 years of cultured limbal epithelial cell therapy: a review of the outcomes. J Cell Biochem 2011;112:993–1002.

[35] Tseng SC, et al. Concept and application of limbal stem cells. Eye 1989;3:141–57.

[36] Fagerholm P, Lagali NS, Merrett K, Jackson WB, Munger R, Liu Y, Polarek JW, Söderqvist M, Griffith M. A biosynthetic alternative to human donor tissue for inducing corneal regeneration: 24-month follow-up of a phase 1 clinical study. Sci Transl Med 2010;2:46ra61.

[37] Koulikovska M, Rafat M, Petrovski G, Veréb Z, Akhtar S, Fagerholm P, Lagali N. Enhanced regeneration of corneal tissue via a bioengineered collagen construct implanted by a nondisruptive surgical technique. Tissue Eng Part A 2015;21:1116–30.

[38] Rafat M, Li F, Fagerholm P, Lagali NS, Watsky MA, Munger R, Matsuura T, Griffith M. PEG-stabilized carbodiimide crosslinked collagen-chitosan hydrogels for corneal tissue engineering. Biomaterials 2008;29:3960–72.

[39] Cieślar-Pobuda A, Rafat M, Knoflach V, Skonieczna M, Hudecki A, Małecki A, Urasińska E, Ghavami S, Łos MJ. Human induced pluripotent stem cell differentiation and direct transdifferentiation into corneal epithelial-like cells. Oncotarget 2016;7:42314–29.

[40] Hayashi R, Ishikawa Y, Sasamoto Y, Katori R, Nomura N, Ichikawa T, Araki S, Soma T, Kawasaki S, Sekiguchi K, et al. Co-ordinated ocular development from human iPS cells and recovery of corneal function. Nature 2016;531:376–80.

[41] Mikhailova A, Ilmarinen T, Ratnayake A, Petrovski G, Uusitalo H, Skottman H, Rafat M. Human pluripotent stem cell-derived limbal epithelial stem cells on bioengineered matrices for corneal reconstruction. Exp Eye Res 2016;146:26–34.

Chapter 13

Successful Biomaterial-Based Artificial Organ—Updates on Artificial Blood Vessels

Marek J. Łos[*,§], Soumya Panigrahi[†], Katarzyna Sielatycka[‡], Catherine Grillon[§]

[*]*Biotechnology Centre, Silesian University of Technology, Gliwice, Poland*

[†]*Case Western Reserve University, Cleveland, OH, United States*

[‡]*Department of Physiology, Faculty of Biology, University of Szczecin, Szczecin, Poland*

[§]*Center for Molecular Biophysics, UPR4301 CNRS CS80054, Orleans Cedex 2, France*

INTRODUCTION

The U.S. Department of Health & Human Services reported in 2017 that the number of patients waiting for an organ or tissue donation in the US is over 121,000 in the donation list, but there were only 14,000 available donors and this gap continues to widen [1]. The increasing demands and paucity of universal tissue donors turned tissue engineering into an emerging field of research and the most useful approach to address this crisis.

Surgical corrections of defects in large and small size arteries and a multitude of cardiac anomalies require trustworthy biocompatible grafts. Maintaining proper anatomical integrity and steady functions as major high shear force blood vessels, coronary arteries, heart valves with expected long-term results require highly specialized reconstruction materials. Any synthetic and/or biogenic and humanized graft with necessary physiological compliance, structural durability, and resistance to bacterial colonization could serve the purpose. Moreover, in the milieu of reconstructive vascular surgery, there is a high demand to develop and use clinical grade grafts that could maintain long-term patency rates. The ideal artificial blood vessel needs to be constructed of viable and compatible tissue, with contractile and secretary properties, physiologically compliant under the shear stress of pulsatile blood flow, and sensitive to chemical stimuli. Laboratory grown synthetic vascular graft should also be immunologically compatible, along with healing—remodeling properties, and having potentials

Stem Cells and Biomaterials for Regenerative Medicine. https://doi.org/10.1016/B978-0-12-812258-7.00013-7

to grow and strengthen particularly if implanted in the children. The other essential criteria to be fulfilled are their easy availability in various formats, diameters, lengths, and with relative ease of storage preparation in addition to easy handling requirements.

The use of tissue engineering methods for constructing long-lasting biocompatible substitutes of pathologic blood vessels is one of the ways to partially fulfilling these requirements, although local inflammatory reactions and immune cell infiltrations are frequently observed, which results in infection, calcification, and graft failure. In an attempt to overcome such issues, experimental procedures using in vitro expanded autologous stem cells from the patient and implantation in a biodegradable synthetic polymer tube for reconstructing an occluded pulmonary artery was used in a 4-year-old patient [2, 3]. Subsequent clinical trials, with similar grafts, where autologous bone marrow derived mononuclear stem cells were implanted, showed excellent results with no graft-related mortality in a mean follow-up of 5.8 years [4]. Long-term monitoring by CT scanning and post mortem histological studies indicated extensive remodeling and graft growth in these patients. Such an "off-the-shelf" decellularized graft would require the development of a large-scale in vitro expansion and culture system for isolating the autologous stem cells from bone marrow and reliably seeding them onto the scaffold [2]. Reportedly, one of such advanced biomimicked tubular grafts suitable of replacement blood vessel was produced as a "proof of principle" in a bioreactor where dermal fibroblasts are trapped in a fibrin gel mold [5] and stimulated to produce a collagen scaffold followed by decellularization to eliminate the need for immunosuppressant, but maintains the cell produced matrix [6, 7].

The lack of an efficient vascular system is a problem that limits the adaptation of artificial organs for clinical use. Solutions such as artificial scaffolds containing proangiogenic factors or creating a network of vessels with the use of 3D printers still do not fully mimic the natural system of capillaries. The efficient vascular system must provide oxygen access to all parts of the organ, taking into account the maximum diffusion range of 100–200 μm. To fully mimic the capillary structure, the smallest capillaries must reach a diameter of less than 10 μm and at the same time allow the flow of red blood cells [8, 9].

Although there are some reports of successful generation and application of lab-grown vascular tissue grafts, maintaining adequate metabolic support by proper vascularization of the constructs is one of the key, major obstacles associated with the long-term outcome [10, 11]. In addition, long-lasting controlled local delivery of bioactive substances opens up additional major technical challenges for manufacturing growth-factor-loaded scaffolds [12, 13]. Another potential alternative to direct delivery of proteins is into collagen scaffolds carrying DNA encoding angiogenesis factors to transfect cells in the extracellular matrix (ECM) acting as bioreactors for the local production of tissue-inductive factors [14]. In the following chapters, we will discuss about the recent developments in production methods, quality controls, and future prospects of artificial blood vessels with potential clinical applications.

SCAFFOLDS, VASCULAR CELLS, AND THEIR INTERACTIONS

Vascular tissue engineering is a robustly accepted and promising approach to eliminate hurdles of using currently available vascular substitutes. Rapid restoration of the blood flow to the distal extremities requires readily available artificial blood vessel grafts to replace irreparably damaged native blood vessels in patients with multiple trauma. A biocompatible vascular conduit would be a reasonable substitute. Currently used patch materials like autologous, allogenic, or xenogenic (glutaraldehyde fixed) have considerable limitations like infection, thrombus formation, calcification, foreign body reaction, and lack of growth potential [10, 11]. Other routinely used synthetic vascular grafts materials polyester and expanded polytetrafluoroethylene (ePTFE) have also similar inherent problems. The chance of biological acceptance increases when a graft with an adherent layer of uninterrupted vascular endothelial cell due to its physiological compliance to the circulating blood minimizing thrombosis and occlusion [6]. Adhesion of functioning endothelial cells to the scaffold primarily regulated by the physical and chemical characteristics of the scaffold materials and specific surface modifications methods can greatly enhance endothelial cell adhesion and proliferation on the inner lining of artificial blood vessels [15].

Vascular tissue engineering—the biological bases: bioabsorbable vascular conduits capable of housing regeneration and growth of physiologically compatible vascular wall may overcome the limitations of contemporary patch repair, which are nonviable and without the capacity of growth, repair, and remodeling. These limitations currently downplay the long-term success of a synthetic vascular graft, initiating graft failures and requirements for multiple surgeries. There are two generally accepted ways to address this problem: (1) maintaining the native vascular functions, namely, the mechanical function, and (2) keeping the tissue-engineered vessel as a functional remodeling template. Therefore, from surgical implantation to complete organization, an ideal vascular graft should reproduce all of the structural and functional aspects of natural blood vessel with anti-inflammatory properties, compliant but resistant to shear stress, suturable, nonleaky, and able to remodel. In addition to the structural aspects, the tissue engineering strategies rely on the ability to allow the transplanted allogeneic or host origin cells with attaching and migrating biologically within the scaffold, and to remodel. Remodeling includes mimicking the mechanical and biological properties as well. Application of the principles of cell biology, material science, and biomedical engineering could eventually create biological substitutes with the ability to restore and maintain normal function in diseased and injury [16, 17]. It has been previously reported that during normal tissue development tissue morphogenesis is influenced by interactions between cells and ECM. While simple polymeric scaffolds that have been used in the past provide architectural support for a neotissue development, they do not adequately mimic the complex interactions between tissue-specific cells and tissue-specific

ECM that promote functional tissue regeneration. Thus, future advances in tissue engineering will depend on the development of novel scaffolding systems that actively modulate cell behaviors to build healthy vascular tissue [11]. In the following sections, we will discuss scaffold materials, their assembly techniques and the potential intermingled cellular components, which are currently being investigated for therapeutic applications.

Selection of the Scaffold Material

Advancement in the development of tissue-engineered blood vessels depends on the fabrication of novel scaffolds, which will actively control the behavior of implanted cells to build healthy durable vascular tissue. In current practice, great saphenous vein is the most commonly used native vessels, for the repair and replacement of small-diameter arteries [18, 19]. Advantages are, it is "ready-to-use" patch that is most biocompatible and highly tolerated by the immune system, although the availability is limited or questionable, depending on the individual patients' clinical condition. Furthermore, in critically injured patients, the physiologic insult incurred during the harvesting of the vein must be weighed against the use of an alternative patching agent. Owing to their excellent mechanical properties and biocompatibility, decellularized arteries, from allogeneic, cadaveric, and xenogeneic sources, are attractive alternative for saphenous venous vascular grafts; however, they are limited to their manufacturing precision, possibility of infections, and calcification [20, 21]. In contrast, vascular scaffolds that provide a temporary structural support and desired shape to the implanted cells/tissue until they produce their own supporting extracellular matrix (ECM) elements and maintain it. Most commonly used matrices are collagen based or of biodegradable polymers [22, 23]. Subsequently patient-derived or allogeneic endothelial cells (EC), smooth muscle cells (SMCs), and fibroblasts can be seeded in the three-dimensional scaffold and cultured in a bioreactor until completely cellularized blood vessel with optimal mechanical properties is created. Bioreactors that mimic the in vivo environment of the vascular cells by producing pulsatile flow are being evaluated and may be useful in this process [22, 24, 25].

Bioengineered Vascular Scaffolds

A method was introduced by Hirai et al. in late 1990s where hybrid tissue of *tunica media* origin was prepared by mixing canine jugular vein smooth muscle cells (SMC) and type I collagen. This cell mix was allowed to grow in the inner surface of a tubular glass mold. In the second step, the luminal surface was seeded with jugular vein endothelial cells and the resulting graft was implanted in canine vein. Interestingly, the graft ruptured shortly after implantation despite being in a low-pressure system. Although the patency rate of the wrapped grafts was 64% at 6 months, such grafts were subsequently wrapped with a Dacron mesh to prevent bulging and tearing. Complete remodeling could be documented by a monolayer of longitudinally oriented ECs on the luminal surface, circumferentially oriented SMCs of the contractile phenotype at the sub endothelial layer, and in-grown fibroblasts throughout the vessel wall, meshes of collagen fibrils in the intercellular space, and sheet-like elastic lamellae in the subendothelial layer. In this system, the differentiated SMCs and functional endothelium were able to produce von Willebrand's factor and prostacyclin. These features made this blood vessel model very attractive to the scientists triggering further developments into the tissue engineering of blood vessels. However, the absence of elastin, the longitudinal (instead of circular) orientation of the SMCs, and the low densities of the SMCs and collagen are thought to be the reason for the poor mechanical properties of the model [26, 27].

Alternatively, **electrospinning** is a method that provides finetuned mechanical properties of the scaffold starting from manufacturing to application. A considerable amount of cellular incompatibility can also be avoided by controlling their production and postproduction processing. Electrospinning technique has the advantage of using both synthetic and natural materials providing high mechanical durability and compliance.

With the possibility of incorporating natural polymers having multiple cell binding sites, an electrospun graft can promote the growth of a continuous layer of EC in the luminal surface and help the growth of proliferation of other component cells in the matrix of the luminal wall. Other practical advantages of using electrospinning technique include a precise control over the composition, dimension, and the alignment of structural fibers affecting porosity, pore size distribution, and the architecture of scaffolds. The aligned nanofibers can be used for orienting cells in a specific direction necessary to provide the anisotropy encountered in certain organs including blood vessels [28].

Impact of Fiber Diameter, Alignment, and Pore Size

In the electrospinning scaffolds fiber diameter, alignment, and pore size incur significant impact on the final functionality of the vascular conduit. Peritoneal incubation generally leads to deposition of a thin fibrous capsule containing macrophages, smooth muscle, or smooth-muscle-like cells on the outer surfaces of the vascular constructs. Interestingly, larger pore size improves the cell infiltration, elongation, and matrix deposition in the scaffold. Under some experimental conditions, small-diameter vascular constructs with a matrix organization mimicking vascular tissue could be produced by peritoneal cells around tube scaffold inserted into the peritoneal cavity [29, 30]. Electrospun fiber diameter impacts the pattern of cell growth within and around the conduit. The extent of cell infiltration into the conduit wall is also modulated by changes in scaffolding characteristics from the luminal to outer surfaces. Because of local variations in the cellular microenvironment, the individual ECM components of the vascular scaffolds in such a system and the surrounding capsule show morphological differences. In addition, the more elongated cells were found within the fibrous capsule, which produced a higher amount of matrix components compared to the rounded cells of the conduit wall. These differences include the presence of porous, hydrolytically degradable PCL fibers within the conduit wall [30].

Ex Vivo Behavior of the Constituent Vascular Cells

The endothelial cell monolayer provides a smooth inner lining of the vascular tree playing a vital role in preventing spontaneous coagulation and in the maintenance of vascular tone in a variety of pathophysiological conditions. Although primary human endothelial cells obtained from fetal umbilical vein to large and small size arteries are often being used to study various aspects of vascular biology, particularly in drug screening and cell-based assays, the possibility of a wide applicability of these calls—both patient derived or allogeneic—is only being explored (https://www.futuremedicine.com/doi/pdf/10.2217/rme.15.73). In the recent years, growing human primary vascular cells in specialized growth factor rich medium is gaining increased research and commercial attention. However, it is challenging to harvest graft recipients' primary vascular tissue and expanding sufficient number of primary cells in vitro for successfully implanting them in vascular scaffolds. Therefore, the use of clinical grade early passage endothelial, adventitial, and aortic smooth muscle cells for repopulating the vascular scaffold is the next best choice provided sufficient immunological tolerance could have been established [31]. Alternatively, the use of ex vivo amplified individual constituting cells from human umbilical vein could be less immunogenic and requires further explorations.

A critical extracellular component of vascular tissues is the abundance of matrix collagen produced by fibroblasts. Fibroblast activation is also a common occurrence after inflammation [32]. Associated deposition of elastic matrix in the vascular constructs is also indicative of local attempt of blood vessel generation. This implies the importance of incorporating elastogenic factors into the artificial vascular conduits for activating elastic matrix generation [33].

Cells recruited to the synthetic vascular conduits within the small-pore pouches are predominately CD68-positive tissue macrophages [34]. The presence of extensive hyaluronic acid (HA) deposits in constructs within small-pore pouches further supports a delayed inflammatory response, since HA accumulation has been observed at sites of both vascular and peritoneal wound healing. The presence of HA itself is not a concern since long-chain HA is known to play an important role in maintaining vascular integrity [34]. The large-pore scaffolds promote a higher healing response for the conduits characterized by increased cellularity, cell infiltration, and presence of a higher number of elongated cells indicating the benefits of a large pore size in the implant [34].

Several strategies have been employed to stimulate reorientation of the endothelial cells (EC), and aortic smooth muscle cells (SMC) in a circumferential direction within the scaffold because in the collagen-based artificial vessels the smooth muscle cells acquire a predominantly longitudinal orientation, which has been partially attributed to the associated poor mechanical strength. To circumvent such issues while attempting to substitute worn out tissues and to potentially regrow parts or whole organs, the use of stem cells with plasticity of accommodating the specific tissue microenvironment is gaining attention [35]. The prospects of using mesenchymal stem cells (MSC) in tissue repair and regenerative medicine also enrich our knowledge in this area. Although the use of stem cells to repopulate the vascular scaffold in a physiologically polarized state is still in the research phase or at the stage of ethical debates, some therapies based on such tissue engineering concepts are already successfully passing clinical trials [36]. Although the issues requiring further studies still hold inadequately answered questions of the optimal isolation methods of required cell types, and also to obtain adequate differentiation into specific cell lines (such as EC, SMC, adipocytes, chondrocytes, and osteoblasts) [37].

Impact of the Biomaterial Constituents

Attempts were made to test in the clinic biodegradable scaffold-based artificial blood vessels seeded with isolated vascular cells from peripheral vein. In such experiments, endothelial cells isolated from peripheral vein were seeded on a reinforced polymer tube. A porous scaffold constructed from degradable copolymer mesh of poly-L-lactide (PLLA) and poly-ε-caprolactone (PCL) was reinforced with polyglycolide (PGA). Such vascular grafts were used as extracardiac conduits to correct single ventricular birth defect, and have been tested in 25 patients for up to 7 years with no evidence of graft occlusion or aneurysm formation [38]. Alternatively synthetic cell free biodegradable scaffolds could be implanted allowing the body to be exploited as a bioreactor and cell source to seed the acellular matrix. Several tissues, including skin, fascia, peritoneum, muscle, small intestine submucosa, and ureter, have been used for this purpose with various outcomes [10].

Engineered vessels showed measurable contractions to pharmacologic agents such as serotonin, endothelin-1, and contained well-differentiated SMCs, as displayed by the expression of calponin and myosin heavy chains. The initial experience with two of these grafts implanted into the saphenous artery of miniature swine was promising. They were functioning for up to 4 weeks, in contrast to two nonpulsed grafts that thrombosed [39]. Prostaglandin F2α was added in the culture medium, which was flowing through the vascular lumen to artificially generate physiologic shear force and pulsatile stretch on the luminal surface to the vessel wall while flowing culture medium with a bioreactor. A copolymer of PGAP4HB was used as the scaffold material because its strength, pliability, and thermoplastic characteristics allowed fabrication into almost any shape using a heat-welding technique. Grafts were sequentially seeded with ovine vascular myofibroblasts and ECs and placed in the pulse duplicator system for up to 4 weeks. Increased cell mass and collagen content, well organized in a layered fashion, was achieved in the pulsed vascular grafts, resulting in supraphysiologic burst pressures and suture retention strength appropriate for surgical implantation within 3 weeks of culture. In addition, the authors suggest that individual cell pools could be created from a person's own umbilical cord for potential future use. This technique would eliminate the need to harvest autologous cells and the problem of immunogenicity of heterogenic cells.

Decellularized Vascular Scaffolds

It has been successfully shown in animal models that decellularized allogeneic or xenogeneic arteries would give very good biomechanical profile for vascular tissue engineering [40]. However, decellularized canine arteries did not appear to perform well when implanted as interposition carotid or coronary artery bypass grafts in dogs, with a high thrombosis rates. The thrombogenicity of decellularized xenogeneic vessels is mainly due to the lack of EC lining of the luminal surface of the vessel. To overcome this problem, reseeding of decellularized porcine aortas with pre-expanded human ECs and myofibroblasts harvested from saphenous vein biopsy specimens was tried with partial success. Apart from demonstrating the feasibility of reseeding decellularized vessels with vascular cells, these experiments also revealed that the in vivo immune response was greatly reduced after such processing. Ex vivo reseeding of decellularized vascular matrix scaffolds could therefore be the answer to thrombogenicity, immunologic rejection, and lack of healing, which are the three main problems of cryopreserved or glutaraldehyde-fixed grafts.

Scaffolds From Natural Polymeric Biomaterials

The natural polymers used in tissue engineering are primarily sugar and protein compounds. They are not only the material that builds the scaffolding, but also the nutrients for the cells. Although they have many advantages, the scaffold fabricated from these naturally occurring polymers is mostly cell friendly and exceedingly biocompatible, they are not the best materials for scaffolding. They are usually unstable in different temperatures and environmental conditions, which makes it difficult to find the right method of scaffolding production. Furthermore, the structures made of these polymers generally have poor mechanical strength. For these reasons, composite materials that are mixtures of natural and synthetic polymers are often used [41, 42].

Collagen is the most abundant ECM protein seen in our body. Several collagen types have been identified. It is the main protein of the connective tissue, responsible for giving the tissue the appropriate strength and mechanical resistance. Collagen is the most popular and well-investigated natural biomaterial [43, 44]. Collagen nanofibers are compatible with a number of cell types, including myoblasts and chondrocytes [44, 45]. In addition to biodegradability and biocompatibility, its advantage is good cell adhesion. In spite of many beneficial features, it is not often used as an independent construction material because it has insufficient mechanical properties (unlike in vivo, in vitro manufacturing does not generate crosslinking between individual fibers) and short degradation time. It is used in combination with other synthetic polymers [41, 45]. Huang et al. studied the blending of type I collagen nanofibers (produced by electrospinning) with poly(ethylene oxide) (PEO). Their results demonstrated that due to a high number of intermolecular interactions between collagen and PEO, the mechanical strength of the nanofiber system was significantly increased [46, 47]. These studies give a chance for an increased role of collagen in tissue engineering.

Chitosan is a chitin derivative and it finds a wide variety of applications as scaffold in tissue engineering due to its low cost, ease of production, antimicrobial activity, and biocompatibility [48]. It is biodegradable, bioresorbable, and it has good adhesion to cells. Chitosan has structural similarity to GAGs and comparable degradability. The poor mechanical property of chitosan-based scaffolds is solved by blending it with synthetic polymers [49], and even with gelatin [50, 51]. In the case of scaffolding, it is mainly used together with other natural polymers, i.e., hyaluronic acid or elastin [52]. Nonwoven or aligned chitosan/PEO (90:10) nanofibers have been developed using the electrospinning technique. These nanofibers possessed structural integrity in water and their cell studies demonstrated enhanced attachment of human osteoblasts and chondrocytes onto the nanofibers. In addition, the cells maintained their characteristic morphology and viability on these nanofibers, thereby indicating good cytocompatibility [46]. Hence, chitosan nanofibers could be a good candidate material for scaffolds in tissue engineering.

Hyaluronic acid is a polysaccharide found in organisms and also present in the synovial fluid. In tissue engineering, it is used for the manufacture of scaffolds for chondrocyte cultures along with synthetic polymers [53]. Um et al. developed nanofibers of hyaluronic acid using the technique of electrospinning and electroblowing (blowing hot air during the process of electrospinning). The authors observed that electrospinning of hyaluronic acid does not allow the consistent production of high-quality nonwoven nanofibers [54]. Therefore, they employed a new technique of electroblowing that was a combination of electrospinning and air flow. In this study, the authors produced nanofibers of hyaluronic acid via electrospinning and by blowing air at 57°C with a 70 ft/h flow rate.

Elastin is one of the major ECM proteins in the arterial wall [55] that confers elastic recoil, resilience, and durability [56]. It is an important autocrine regulator to SMC and EC activity, inhibiting migration and proliferation of SMCs and enhancing attachment and proliferation of ECs. Elastin, as a coating of vascular devices (made of ePTFE [57] PET, a copolymer of ePTFE and polyethylene, and a polycarbonate polyurethane), demonstrated low thrombogenicity with reduced platelet adhesion and activation [57, 58].

Another natural biomaterial that has been well studied is *gelatin*. Zhang et al. developed gelatin/PCL composite fibrous scaffolds using the electrospinning technique. Their study indicated that the composite nanofibers have improved mechanical strength and wettability compared with gelatin or PCL alone [59].

Silk is another potential natural biomaterial for nanofibrated scaffolds. Min et al. described the in vitro cytocompatibility of silk nanofibers with keratinocytes and fibroblasts. It is a suitable material for use in scaffold technology because it is compatible with cytokines, and has predominantly fiber diameters and high porosity [60].

Polymeric Scaffolds

The relatively weak mechanical properties (when manufactured in vitro) limit the usage of natural polymers for vascular graft application. Synthetic polymers have better mechanical resistance and durability in various conditions. For these reasons, they are commonly used to manufacture scaffoldings and they represent the largest class of biomaterials [32]. In the case of small diameter grafts, the ability to withstand the physiological pressure without bursting is a necessity. Most of the naturally derived scaffolds fail to meet that requirement. Thus, it is reasonable to apply a synthetic macromolecule-derived scaffolds as most of its properties can be tailor made during its synthesis.

Polycaprolactone (PCL) is a biodegradable and nonpolar aliphatic polyester obtained from petrochemical raw materials. It is obtained as a result of caprolactone polymerization. It is characterized by fairly good plasticity and mechanical strength. The material is biodegradable. It is hydrolyzed to 6-hydroxycaproic acid, which is converted into the citric acid cycle and is then excreted from the body. Polycaprolactone has a longer degradation time than, e.g., polylactide (over 2 years). For this reason, it is used as a long-term implant or as a composite with other polymers, both natural and synthetic. PCL is also widely used in the manufacture of surgical screws and screws as well as disposable items [61].

Polylactide (PLA) is classified as biocompatible, biodegradable, nontoxic, and is FDA-approved for human implantation [62]. PLA is biodegradable polyester. Degradation of PLA in vivo is facilitated through a hydrolytic splitting of the ester bonds that leads to the formation of the natural metabolite lactic acid and glycolic acid. A possibility for the regulation of the degradation rate is the copolymerization of PLA with PGA [41].

Polyglycolide (PGA) is the simplest aliphatic polyester made up of glycolic acid units. Also, PGA is classified as biocompatible and nontoxic and is FDA-approved for human implantation. It is a polymer derived from renewable sources. It is characterized by a very short time of hydrolytic degradation (several weeks). Rapid degradation is the result of high hygroscopicity of the polymer—water penetrates between the chains, breaks them, which facilitates the transport of hydrolytic enzymes. PGA is used, among others in medicine, for the production of quickly decaying surgical threads and needles [63]. Sodian et al. reported on the production of a tricuspid heart valve made of a PGA structure that unfortunately demonstrated insufficient structural integrity under physiological flow and pressure conditions [64].

Poly (lactide-co-glycolide) (PLGA) is a copolymer of lactic and glycolic acid. Depending on the ratio of the two compounds used for the polymerization, a polymer with suitable properties is obtained. Advantage of PLGA is the ease of controlling the degradation time depending on its composition. Due to the addition of lactic acid, degradation of the polymer occurs more slowly than the compound made only of glycolic acid [62].

METHODS OF GENERATING BIOCOMPATIBLE VASCULAR SCAFFOLDS

The development of nanofibers significantly increased the possibilities of scaffolding production that mimics the tissue architecture at the nanoscale. The development of appropriate scaffoldings is a milestone in the field of tissue engineering. Currently, three techniques are available for the synthesis of nanofibers: electrospinning [65], self-assembly [66], and phase separation [67]. The most frequently tested technique and showing the most promising results in the field of tissue engineering applications is electrospinning because there is a wide range of natural and synthetic biomaterials that contribute to increasing the scope of development of nanofibrid scaffolds. Three-dimensional synthetic biodegradable scaffolds designed using nanofibers serve as an excellent framework for cell adhesion, proliferation, and differentiation. Nanofibers, regardless of their synthesis method, are used as scaffolds for the engineering of musculoskeletal tissue, skin tissue engineering, vascular tissue engineering, neural tissue engineering and as carriers for controlled drug delivery, proteins, and DNA [46].

Auto/Allo/Genografts

One of the most promising techniques for tissue regeneration is tissue repair by autologous cell or tissue transplantation. However, autografts are associated with limitations such as donor site morbidity and limited availability. Another solution is allograft, which are tissue taken from another individual of the same species, even though they may elicit an immune response, and also carry a risk of disease transmission. Tissue engineering has proven to be a great approach to repair or to regenerate damaged tissue, with the ability to circumvent all limitations of autologous and allogeneic tissue repair.

Electrospinning in Two and Three Dimensions

Electrospinning is the most promising approach for large-scale production of nanofibers, and the fiber diameter can be adjusted from nanometers to micrometers [68, 69]. Functional bioscaffolds need to meet several requirements: (i) the scaffolding should have a high degree of porosity, with an appropriate pore size distribution, (ii) large surface area (for cell adhesion), (iii) biodegradability is required and the rate of degradation corresponds to the speed of tissue formation, (iv) the scaffold must have the required structural integrity to prevent the scaffolding from collapsing when new vessels are created, with appropriate mechanical properties, (v) the scaffold should be nontoxic to cells and biocompatible, positively interact with cells to promote adhesion, proliferation, migration, and diversified cell function. Among the assessed biomedical assessments, electrospatial scaffolds of nanofibers show excellent results in the binding, proliferation, and penetration of cells, both in vitro and in vivo. The scaffold structure produced by the electrospinning process depends largely on the quality of the fibers. Material selection and parameter setting are among many factors that contribute to the quality of the finest fibers that ultimately determine the efficiency of tissue engineering scaffolds. The advantage of this technique is the ability to create a network of nanofibers with high porosity and surface area [70]. Cells behave differently when cultured in 3D compared to traditional 2D cultures. The 3D environment influences the diffusion and adhesion of proteins, growth factors, and enzymes, which ensures cell viability and can influence function [46]. Culturing cells in 3D alters the mechanical signaling from those provided in 2D, thus affecting cell-receptor ligation, intercellular signaling, and cellular migration [71]. Cells cultured on electrospun scaffold often adhere to the surface of the scaffold with minimum penetration. This may be due to a combination of small interfiber distances, fiber diameters, size of the cells, and the chemical and interfacial properties of the fibers. While in these instances the cells do not migrate into the scaffold, it is still uniquely different from a 2D planar surface, as the cells receive nutrients and growth cues three-dimensionally [70].

Silk-Based Biomaterial

Silk fibroin (SF) is an advanced naturally derived biopolymer that offers several features including unique mechanical properties, versatile processability to different formats, proper biocompatibility, very low immunogenicity, and adaptable biodegradability, which make it a perfect choice for vascular scaffolds [72, 73]. Also, it has been revealed that the SF is more biocompatible than other commonly used polymeric degradable biomaterials such as PLA, PGA, and collagen. Silk can be chemically modified through amino acid side chains to alter surface properties or to immobilize cellular growth factors. Molecular engineering of silk sequences has been used to modify silks with specific features, such as cell recognition or mineralization. According to the mentioned parameters, SF is a promising option among the polymeric materials for scaffolding in the regeneration of vascular grafts. Hence, various methods have been employed for the fabrication of small-diameter silk-based vascular grafts such as electrospinning, gel spinning, dipping, and bilayer grafts composed of SF or composites of SF with other polymers, which have been appraised in in vitro and in vivo settings [72].

Silk-Based Electrospun Tubular Scaffolds

Responding to the demand for vascular skeleton produced by the electroconviation method, with the features of biocompatibility, degradation, and general functionality, it was decided to investigate new protein polymers, among others Bombyx silk fibroin with morisilkworm. This material proved to be suitable for a wide range of biomaterials and tissue engineering needs, thanks to the unique combination of impressive mechanical properties, slow degradation, and biocompatibility [74]. Electrospun silk scaffolds have promising mechanical characteristics for vascular grafts, including the ability to withstand arterial pressures and behave in a mechanically similar mode to native vessels. These systems also support vascular endothelial and smooth muscle cells, suggesting future potential for these protein systems in vessel applications. Scaffold properties may be further improved through fiber alignment and increased control of the electrospinning parameters. Studies with the use of electron silk are presented in the work from Soffer et al., where human aortic endothelial cells and smooth coronary artery smooth muscle cells attached, spread, and grew on the electrospun silk mats [74]. The endothelial cells remained on the surface of the electroporous mats due to small size pores in the nonwoven matrices, while some of the smooth muscle cells migrated under the mats. The ability to promote vascular-important cells suggests that electron silk tubes can be run in the context of further detailed studies for vascular graft materials [74].

Electrospun Silk-Based Fibers

Silk fibroin has been used to generate fibrous silk membranes from reprocessed native silk fibers or by electrospinning [75, 76]. Electrospun fibrous membranes are of interest as biomaterials because the increased surface area is favorable for cell adhesion [46, 77]. 3-D constructs of nanofibers have been used as blood vessel grafts and nerve guides [78]. Combination of type I calf skin collagen and chitin with silkworm (*Bombyx mori*) fibroin, during the electrospin process resulted in fibers supporting cell adhesion, the proliferation of human epidermal keratinocytes (NHEK), and fibroblasts (NHEF) in vitro. Interestingly, electrospun fibers composed of unmixed *B. mori* fibroin and type I calf skin collagen or *B. mori* fibroin and chitin (prepared by electrospinning from two syringes simultaneously and subsequently crosslinked with glutaraldehyde) showed higher levels of cell adhesion and proliferation for NHEK cells in vitro [79]. The mechanical properties of electrospun SF scaffolds were enhanced by uniformly dispersing hydroxyapatite (HAp) nanoparticles within the SF nanofibers [79].

3D Printing

The production of blood vessels causes many difficulties mainly due to the required physical properties, biocompatibility requirement, and also the tubular shape [80]. The 3D printing is already used to produce bone and dental implants. Blood vessels are "printed" using a 3D printer that uses agarose as a material, which is a compound of natural origin.

The scaffolding formed by the agarose fibers is coated with a hydrogel and the structure thus obtained is reinforced with photocrosslinks [81]. Unfortunately, the

3D printing method has its limitations. It is impossible to obtain very small vessels, and capillaries [80].

Recently, another method of blood-vessel printing has been developed. It involves vessel printing on a predetermined tubular support that corresponds to the expected inner diameter of the vessel (http://www.xinhuanet.com/english/2015-10/25/c_134748306.htm). The suitability of the technology for the production of clinically compatible human vessels is under evaluation.

STRUCTURAL REMODELING OF THE SCAFFOLDS DURING TISSUE DEVELOPMENT

ECM is a highly dynamic structure that is present in all tissues and continuously undergoes controlled remodeling. ECM interacts with cells to regulate diverse functions, including proliferation, migration, and differentiation [82]. This process involves quantitative and qualitative changes in the ECM, mediated by specific enzymes that are responsible for ECM degradation, such as metalloproteinases. The extracellular matrix is a meshwork of both structural and functional proteins assembled in unique tissue-specific architectures. The ECM both provides the mechanical framework for each tissue and organ and is a substrate for cell signaling. The ECM is highly dynamic, and cells both receive signals from the ECM and contribute to its content and organization. This process of "dynamic reciprocity" is the key to tissue development and for homeostasis. Based upon these important functions, ECM-based materials have been used in a wide variety of tissue engineering and regenerative medicine approaches to tissue reconstruction. It has been demonstrated that ECM-based materials, when appropriately prepared, can act as inductive templates for constructive remodeling. Such materials act as templates for the induction of de novo functional, site-appropriate, tissue formation. The ECM provides a naturally occurring substrate for cell viability and growth. As applications of ECM scaffolds in tissue engineering and regenerative medicine move toward the reconstruction of increasingly complex tissue structures and even whole organs, it is important to understand the mechanisms by which ECM scaffolds promote constructive remodeling [83].

Scaffold Repopulation by Surrounding Vascular Endothelium

Process of ECM scaffold-mediated remodeling is based on the fact that, in each case, an initially acellular scaffold is populated by host cells and undergoes a remodeling process resulting in replacement of the material with functional, site-appropriate host tissue. The outcomes are distinctly different from the default mammalian response to tissue injury, which commonly results in scar tissue formation with a loss of function. The mechanisms by which such scaffold materials are capable of constructive remodeling in diverse applications is not yet fully elucidated; however, it includes the exposure to mechanical forces (shear stress from blood flow), modulation of the host immune response, and degradation of the scaffold material with concurrent new tissue development. In the absence of any of these key factors, constructive remodeling is not observed and outcomes are undesirable [83].

Impact of Matrix Turnover: ECM Production and Scaffold Degradation

The extracellular matrix provides a three-dimensional structural support occupying the space between cells, is a substrate for cell migration, and is a transmitter of biomechanical forces. Scaffolds should contain as many physiological characteristics of the natural extracellular matrix as possible. The physical properties of the ECM, such as rigidity, porosity, insolubility, and topography that derive from composition of the matrix largely determine the mechanical behavior of each individual tissue as well the behavior of the cells which reside within. Numerous studies have demonstrated the potential effects of ECM ultrastructure and mechanics upon cell behavior, migration, and differentiation [83, 84]. The composition of the ECM varies greatly from tissue to tissue, and in some cases within a given tissue. For example, articular cartilage contains large amounts of collagen II and glycosaminoglycans, which are specifically tailored to accommodate high water content and allow resistance to and recovery from compressive deformation. In contrast, tissues such as tendon contain much higher amounts of collagen I, and an organization designed to resist tensile loading. Mainly interstitial cells and endothelial cells are used in cardiovascular tissue engineering. The scaffold may be recognized as a foreign body and, therefore, a target for thrombocytes and also for immune cells. Hence, EC layer shields the scaffold material and reduces both thrombogenicity and immunogenicity, the latter being the reason for the prostheses' degradation. The mechanisms of in vivo degradation of ECM scaffolds are complex and include both cellular and enzymatic pathways. The process is mediated by inflammatory cells, such as macrophages, which produce oxidants as well as proteolytic enzymes that aid in the degradation of the matrix. The major families of proteinases, which are responsible for degradation of the ECM, are the MMP and metalloproteinase with thrombospondin motif families (ADAMTS) [85]. There are 23 identified MMP members [86] and 19 ADAMTS members. Group Valentin et al. have showed, utilizing ^{14}C-labeled ECM scaffolds, that peripheral blood monocytes are required for the early and rapid degradation of ECM scaffolds, and that crosslinked ECM scaffolds are resistant to macrophage-mediated degradation [87]. ECM scaffolds can also be degraded in vitro by chemical and physical methods.

FUTURE PERSPECTIVES AND CONCLUSIONS

Cardiovascular diseases rank among the top causes of death. Thus, there is a steady increase in demand for vascular prosthesis for blood vessel replacement surgeries and research development of improved materials for diseased tissue replacement. Most implants that are in direct contact with the bloodstream show thrombogenicity and are very susceptible to infections [88]. Small caliber vessels are prone to stenosis due to thrombus formation or intimal hyperplasia. The challenge for producing vascular grafts is to engineer vascular replacements that can withstand pulsation; it means they have adequate shape recovery, and are able to withstand the high pressure and flow rate of the blood stream [89]. A blood vessel, being a dynamic tissue, relies much on highly oriented extracellular matrix. Scaffold architecture thus should mimic the native blood vessel ECM and should adequately allow vascular regeneration in vivo. Scientists are investigating a variety of polymeric biomaterials that could meet the ideal requirements. They range from naturally occurring polymer to synthetic ones, and even decellularized scaffolds. The critical challenges include the optimization of vessel material so that it can withstand systemic arterial pressure and have the capacity to remodel. The surface modification of the scaffold materials has shown improvement of many of the properties and it is an attractive area of research [51]. The future of vascular regeneration relies on bringing all the aspects together to generate a highly graft with a cost effectiveness. Pioneering research interests in the mentioned areas will definitely help to elucidate the true potential of tissue engineered small diameter vascular grafts and enable this technology to reach clinical application.

REFERENCES

[1] U.S. Government Information on Organ Donation and Transplantation. https://www.organdonor.gov/statistics-stories/statistics.html.

[2] Shin'oka T, Imai Y, Ikada Y. Transplantation of a tissue-engineered pulmonary artery. N Engl J Med 2001;344(7):532–3.

[3] Li Y, Meng H, Liu Y, Lee BP. Fibrin gel as an injectable biodegradable scaffold and cell carrier for tissue engineering. ScientificWorldJournal 2015;2015:685690.

[4] Shin'oka T. Clinical results of tissue-engineered vascular autografts seeded with autologous bone marrow cells. Nihon Geka Gakkai Zasshi 2004;105(8):459–63.

[5] Jockenhoevel S. Fibrin gel—advantages of a new scaffold in cardiovascular tissue engineering. Eur J Cardiothorac Surg 2001;19(4):424–30.

[6] Jockenhoevel S, Chalabi K, Sachweh JS, Groesdonk HV, Demircan L, Grossmann M, et al. Tissue engineering: complete autologous valve conduit—a new moulding technique. Thorac Cardiovasc Surg 2001;49(5):287–90.

[7] Zhou H, Xu HH. The fast release of stem cells from alginate-fibrin microbeads in injectable scaffolds for bone tissue engineering. Biomaterials 2011;32(30):7503–13.

[8] Lewis T. Could 3D printing solve the organ transplant shortage? 2017. https://www.theguardian.com/technology/2017/jul/30/will-3d-printing-solve-the-organ-transplant-shortage.

[9] Lovett M, Lee K, Edwards A, Kaplan DL. Vascularization strategies for tissue engineering. Tissue Eng Part B Rev 2009;353–70.

[10] Kakisis JD, Liapis CD, Breuer C, Sumpio BE. Artificial blood vessel: the Holy Grail of peripheral vascular surgery. J Vasc Surg 2005;41(2):349–54.

[11] Jiang B, Brey EM. Formation of stable vascular networks in engineered tissues. In: Regenerative medicine and tissue engineering—cells and biomaterials. IntechOpen. 2011. p. 477–502. https://doi.org/10.5772/837. https://api.intechopen.com/chapter/pdf-preview/19035.
[12] Zhao H, Wu J, Zhu J, et al. Research advances in tissue engineering materials for sustained release of growth factors. BioMed Research International 2015;2015:808202. https://doi.org/10.1155/2015/808202.
[13] Chu H, Wang Y. Therapeutic angiogenesis: controlled delivery of angiogenic factors. Ther Deliv 2012;3(6):693–714.
[14] Mao Z, et al. Enhanced angiogenesis of porous collagen scaffolds by incorporation of TMC/DNA complexes encoding vascular endothelial growth factor. Acta Biomater 2009;5(8):2983–94.
[15] Ahmed F, et al. Engineering interaction between bone marrow derived endothelial cells and electrospun surfaces for artificial vascular graft applications. Biomacromolecules 2014;15(4):1276–87.
[16] Ravi S, Chaikof EL. Biomaterials for vascular tissue engineering. Regen Med 2010;5(1):107–20.
[17] Shum-Tim D, et al. Tissue engineering of autologous aorta using a new biodegradable polymer. Ann Thorac Surg 1999;68(6):2298–304 [discussion 2305].
[18] Fox CJ, et al. Contemporary management of wartime vascular trauma. J Vasc Surg 2005;41(4):638–44.
[19] Muto A, et al. Patches for carotid artery endarterectomy: current materials and prospects. J Vasc Surg 2009;50(1):206–13.
[20] Quint C, et al. Allogeneic human tissue-engineered blood vessel. J Vasc Surg 2012;55(3):790–8.
[21] Steinhoff G, et al. Tissue engineering of pulmonary heart valves on allogenic acellular matrix conduits: *in vivo* restoration of valve tissue. Circulation 2000;102(19 Suppl. 3):III50–5.
[22] Atala A. Recent developments in tissue engineering and regenerative medicine. Curr Opin Pediatr 2006;18(2):167–71.
[23] Chen C, Dubin R, Kim MC. Emerging trends and new developments in regenerative medicine: a scientometric update (2000—2014). Expert Opin Biol Ther 2014;14(9):1295–317.
[24] Atala A. Engineering organs. Curr Opin Biotechnol 2009;20(5):575–92.
[25] Orlando G, et al. Regenerative medicine and organ transplantation: past, present, and future. Transplantation 2011;91(12):1310–7.
[26] Hirai J, et al. Highly oriented, tubular hybrid vascular tissue for a low pressure circulatory system. ASAIO J 1994;40(3):M383–8.
[27] Hirai J, Matsuda T. Venous reconstruction using hybrid vascular tissue composed of vascular cells and collagen: tissue regeneration process. Cell Transplant 1996;5(1):93–105.
[28] Hasan A, et al. Electrospun scaffolds for tissue engineering of vascular grafts. Acta Biomater 2014;10(1):11–25.
[29] Campbell JH, Efendy JL, Campbell GR. Novel vascular graft grown within recipient's own peritoneal cavity. Circ Res 1999;85(12):1173–8.
[30] Campbell GR, Campbell JH. Vascular smooth muscle and arterial calcification. Z Kardiol 2000;89(Suppl 2):54–62.
[31] Watt SM, et al. The angiogenic properties of mesenchymal stem/stromal cells and their therapeutic potential. Br Med Bull 2013;108:25–53.
[32] Kim L, et al. Overexpression of transforming growth factor-beta 1 in the valvular fibrosis of chronic rheumatic heart disease. J Korean Med Sci 2008;23(1):41–8.
[33] Bashur CA, Eagleton MJ, Ramamurthi A. Impact of electrospun conduit fiber diameter and enclosing pouch pore size on vascular constructs grown within rat peritoneal cavities. Tissue Eng Part A 2013;19(7–8):809–23.

[34] Bashur CA, Rao RR, Ramamurthi A. Perspectives on stem cell-based elastic matrix regenerative therapies for abdominal aortic aneurysms. Stem Cells Transl Med 2013;2(6):401–8.
[35] Roemeling-van Rhijn M, Weimar W, Hoogduijn MJ. Mesenchymal stem cells: application for solid-organ transplantation. Curr Opin Organ Transplant 2012;17(1):55–62.
[36] Horch RE, et al. Ethical issues in cellular and molecular medicine and tissue engineering. J Cell Mol Med 2008;12(5B):1785–93.
[37] Atala A. Regenerative medicine strategies. J Pediatr Surg 2012;47(1):17–28.
[38] Hibino N, et al. Late-term results of tissue-engineered vascular grafts in humans. J Thorac Cardiovasc Surg 2010;139(2):431–6. 436 e1–2.
[39] Hoerstrup SP, et al. Tissue engineering of small caliber vascular grafts. Eur J Cardiothorac Surg 2001;20(1):164–9.
[40] Pashneh-Tala S, MacNeil S, Claeyssens F. The tissue-engineered vascular graft-past, present, and future. Tissue Eng Part B Rev 2016;22(1):68–100.
[41] Asti A, Gioglio L. Natural and synthetic biodegradable polymers: different scaffolds for cell expansion and tissue formation. Int J Artif Organs 2014;37(3):187–205.
[42] Guo B, Ma PX. Conducting polymers for tissue engineering. Biomacromolecules 2018;19(6):1764–82.
[43] Mano J, et al. Natural origin biodegradable systems in tissue engineering and regenerative medicine: present status and some moving trends. J R Soc Interface 2007;999–1030.
[44] Matthews JA, et al. Electrospinning of collagen nanofibers. Biomacromolecules 2002;3(2):232–8.
[45] Gersbach CA, et al. Runx2/Cbfa1-genetically engineered skeletal myoblasts mineralize collagen scaffolds *in vitro*. Biotechnol Bioeng 2004;88(3):369–78.
[46] Vasita R, Katti DS. Nanofibers and their applications in tissue engineering. Int J Nanomedicine 2006;1(1):15–30.
[47] Huang L, et al. Engineered collagen-PEO nanofibers and fabrics. J Biomater Sci Polym Ed 2001;12(9):979–93.
[48] Elieh-Ali-Komi D, Hamblin MR. Chitin and chitosan: production and application of versatile biomedical nanomaterials. Int J Adv Res (Indore) 2016;4(3):411–27.
[49] Zhang M, et al. Properties and biocompatibility of chitosan films modified by blending with PEG. Biomaterials 2002;23(13):2641–8.
[50] Mao JS, et al. Structure and properties of bilayer chitosan-gelatin scaffolds. Biomaterials 2003;24(6):1067–74.
[51] Thottappillil N, Nair PD. Scaffolds in vascular regeneration: current status. Vasc Health Risk Manag 2015;79–91.
[52] Sell SA, et al. The use of natural polymers in tissue engineering: a focus on electrospun extracellular matrix analogues. Polymer 2010;2(4):522–53.
[53] Collins MN, Birkinshaw C. Hyaluronic acid based scaffolds for tissue engineering—a review. Carbohydr Polym 2013;92(2):1262–79.
[54] Um IC, et al. Electro-spinning and electro-blowing of hyaluronic acid. Biomacromolecules 2004;5(4):1428–36.
[55] Xu J, Shi GP. Vascular wall extracellular matrix proteins and vascular diseases. Biochim Biophys Acta 2014;1842(11):2106–19.
[56] Liang D, Hsiao BS, Chu B. Functional electrospun nanofibrous scaffolds for biomedical applications. Adv Drug Deliv Rev 2007;59(14):1392–412.
[57] Jordan SW, et al. The effect of a recombinant elastin-mimetic coating of an ePTFE prosthesis on acute thrombogenicity in a baboon arteriovenous shunt. Biomaterials 2007;28(6):1191–7.

[58] Zavan B, et al. Neoarteries grown *in vivo* using a tissue-engineered hyaluronan-based scaffold. FASEB J 2008;22(8):2853–61.
[59] Zhang Y, et al. Electrospinning of gelatin fibers and gelatin/PCL composite fibrous scaffolds. J Biomed Mater Res B Appl Biomater 2005;72(1):156–65.
[60] Park YR, et al. Three-dimensional electrospun silk-fibroin nanofiber for skin tissue engineering. Int J Biol Macromol 2016;93(Pt B):1567–74.
[61] Siddiqui N, et al. PCL-based composite scaffold matrices for tissue engineering applications. Mol Biotechnol 2018.
[62] Makadia HK, Siegel SJ. Poly lactic-co-glycolic acid (PLGA) as biodegradable controlled drug delivery carrier. Polymers (Basel) 2011;3(3):1377–97.
[63] Sekiya N, et al. Efficacy of a poly glycolic acid (PGA)/collagen composite nanofibre scaffold on cell migration and neovascularisation *in vivo* skin defect model. J Plast Surg Hand Surg 2013;47(6):498–502.
[64] Sodian R, et al. Evaluation of biodegradable, three-dimensional matrices for tissue engineering of heart valves. ASAIO J 2000;46(1):107–10.
[65] Dzenis Y. Material science. Spinning continuous fibers for nanotechnology. Science 2004;304(5679):1917–9.
[66] Hong Y, et al. Effect of amino acid sequence and pH on nanofiber formation of self-assembling peptides EAK16-II and EAK16-IV. Biomacromolecules 2003;4(5):1433–42.
[67] Ma PX, Zhang R. Synthetic nano-scale fibrous extracellular matrix. J Biomed Mater Res 1999;46(1):60–72.
[68] Shi X, et al. Electrospinning of nanofibers and their applications for energy devices. J Nanomater 2015;16(1):122.
[69] Jian FJ, Wang X, Tong LT. Functional Applications of Electrospun Nanofibers. InTech – Open Access Publisher. 2011;14:287–326.
[70] O'Brien FJ. Biomaterials & scaffolds for tissue engineering. Mater Today 2011;14(3):88–95.
[71] Rubashkin M, Ou G, Weaver VM. Deconstructing signaling in three dimensions. Biochemistry 2014;53(13):2078–90.
[72] Zamani MZ, Khafaji M, Naji M, Vossoughi M, Alemzadeh I, Haghighipour N. A biomimetic heparinized composite silk-based vascular scaffold with sustained antithrombogenicity. Sci Rep 2017;1(7):4455.
[73] Vepari C, Kaplan DL. Silk as a biomaterial. Prog Polym Sci 2007;32(8–9):991–1007.
[74] Soffer L, et al. Silk-based electrospun tubular scaffolds for tissue-engineered vascular grafts. J Biomater Sci Polym Ed 2008;19(5):653–64.
[75] Daniele MA, et al. Microfluidic strategies for design and assembly of microfibers and nanofibers with tissue engineering and regenerative medicine applications. Adv Healthc Mater 2015;4(1):11–28.
[76] Kundu B, et al. Silk fibroin biomaterials for tissue regenerations. Adv Drug Deliv Rev 2013;65(4):457–70.
[77] Wang K, et al. Facile strategy to generate aligned polymer nanofibers: effects on cell adhesion. ACS Appl Mater Interfaces 2018;10(2):1566–74.
[78] Lovett ML, et al. Gel spinning of silk tubes for tissue engineering. Biomaterials 2008;29(35):4650–7.
[79] Li G, et al. Silk-based biomaterials in biomedical textiles and fiber-based implants. Adv Healthc Mater 2015;4(8):1134–51.
[80] Bertassoni LE, Cecconi M, Manoharan V, Nikkhah M, Hjortnaes J, Cristino AL, et al. Hydrogel bioprinted microchannel networks for vascularization of tissue engineering constructs. Lab Chip 2014;14(13):2202–11.

[81] 3D Printing Polymers with Supramolecular Functionality for Biological Applications; 2018.
[82] Bonnans C, Chou J, Werb Z. Remodelling the extracellular matrix in development and disease. Nat Rev Mol Cell Biol 2014;15(12):786–801.
[83] Brown BN, Badylak SF. Extracellular matrix as an inductive scaffold for functional tissue reconstruction. Transl Res 2014;163(4):268–85.
[84] Reilly GC, Engler AJ. Intrinsic extracellular matrix properties regulate stem cell differentiation. J Biomech 2010;43(1):55–62.
[85] Cawston TE, Young DA. Proteinases involved in matrix turnover during cartilage and bone breakdown. Cell Tissue Res 2010;339(1):221–35.
[86] Page-McCaw A, Ewald AJ, Werb Z. Matrix metalloproteinases and the regulation of tissue remodelling. Nat Rev Mol Cell Biol 2007;8(3):221–33.
[87] Valentin JE, et al. Macrophage participation in the degradation and remodeling of extracellular matrix scaffolds. Tissue Eng Part A 2009;15(7):1687–94.
[88] Ricklefs M, Korossis S, Haverich A, Schilling T. Polymeric Scaffolds for Bioartificial Cardiovascular Prostheses. InTech – Open Access Publisher 2017;12:267–91.
[89] Nisbet DR, Forsythe JS, Shen W, Finkelstein DI, Horne MK. Review paper: a review of the cellular response on electrospun nanofibers for tissue engineering. J Biomater Appl 2009;24(1):7–29.

Index

Note: Page numbers followed by *f* indicate figures and *t* indicate tables.

MIX
Papier | Fördert gute Waldnutzung
FSC® C083411

Kolibri 360 GmbH
Ferdinand-Jühlke-Straße 7
99095 Erfurt, Deutschland
produktsicherheit@kolibri360.de